Jersey Cattle: Their Feeding and Management

by English Jersey Cattle Club

with an introduction by Jackson Chambers

This work contains material that was originally published in 1898.

This publication is within the Public Domain.

This edition is reprinted for educational purposes
and in accordance with all applicable Federal Laws.

Introduction Copyright 2017 by Jackson Chambers

Self Reliance Books

Get more historic titles on animal and stock breeding, gardening and old fashioned skills by visiting us at:

http://selfreliancebooks.blogspot.com/

Introduction

I am pleased to present another title in the "Cattle" series.

The work is in the Public Domain and is re-printed here in accordance with Federal Laws.

As with all reprinted books of this age that are intended to perfectly reproduce the original edition, considerable pains and effort had to be undertaken to correct fading and sometimes outright damage to existing proofs of this title. At times, this task is quite monumental, requiring an almost total "rebuilding" of some pages from digital proofs of multiple copies. Despite this, imperfections still sometimes exist in the final proof and may detract from the visual appearance of the text.

I hope you enjoy reading this book as much as I enjoyed making it available to readers again.

Jackson Chambers

This handbook on the Feeding and General Management of Jersey Cattle, based on information received from Members of the English Jersey Cattle Society, has been revised by a Committee, consisting of the following Members of Council:—

        The Hon. ALEX. E. PARKER,
        The Rev. SYDNEY H. WILLIAMS,
        Mr. RICHARDSON CARR,
        Mr. JOHN FREDERICK HALL,
        Mr. ERNEST MATHEWS, *Editor*.

7, Princes Street, Hanover Square, London, W.
  November, 1898.

PREFACE.

SEVENTEEN years ago there appeared in the *Journal of the Royal Agricultural Society of England* (vol. xvii., 2nd series, part 1, 1881) an article on "Jersey Cattle and their Management," by Mr. John Thornton. This article, which was reprinted in the second volume of the English Jersey Herd Book, gave a short history of Jerseys, and the practice adopted by many leading breeders in England in the feeding and management of their cattle. At that time few breeders kept the Jersey cow for profit. Her elegant appearance attracted notice, and the rich quality of her milk, cream and butter, made the Jersey popular with those who had once tried her; but her remarkable merits as a dairy cow had not then been subjected to public proof, consequently she was not generally to be found in the herds of farmers or of dairymen.

As the number of Members of the English Jersey Cattle Society increased, the entries of Jerseys at the various Shows became so numerous that the leading Societies gave separate classes for them. In 1886, the English Jersey Cattle Society first offered special prizes for those Jersey cows, which yielded in the showyard the largest quantity of butter, by the practical test of the churn. These classes, initiated by Mr. J. F. Hall, who conducted the first tests, are now known as butter test competitions.

During the last few years, the tests at the Bath and West of England Show, the Tring Agricultural Society's Show, and the London Dairy Show have been open to all breeds of cattle. The results prove that, when live-weight, period of lactation, and dairy produce, both in quality and quantity, are taken into consideration, there is no cow which can compare with the Jersey as a butter cow.

The Jersey does not pretend to be a butcher's animal; independently of the cost of feeding, the yellow colour of her fat is objectionable, although the beef is said to be excellent. A cow predisposed to fatten is rarely a good dairy animal, for if she is gaining flesh her milk yield suffers. A good cow should pay for herself many times over in her dairy produce; for although it has been said that, at the present low price of foreign butter, "butter-making would break the Bank of England," yet there is a demand for genuine English butter at a better price, and to meet this demand the Jersey cow is undoubtedly the best animal.

From an article by Mr. J. F. Hall on "Butter Dairying and Butter Cows, British and Foreign," published in the *Bath and West of England Journal*, vol. vii. (4th series, 1896-97), it would appear that the quantity of milk required to make 1 lb. of butter in Denmark and Sweden is about ten quarts (or[1] 25 lbs.), and the average quantity in England of breeds, other than Channel Islands, is about twelve quarts (or 30 lbs.). The tests in England and America of selected animals show that it takes about seven quarts (or 17·5 lbs.), of Jersey milk to make 1 lb. of butter, consequently butter should be made cheaper from Jersey cows than from cows of any other breed.

To make good butter judicious feeding is necessary. Of late years much attention has been devoted to the investigation of the efficiency of particular foods, the balancing of the dietary, and the avoidance of waste. Upon the management of cows at calving time, and the treatment of milk-fever, abortion, and other diseases, a good deal of experience has been accumulating. With a view to obtaining information on these and many other points of almost equal importance, the Council of the English Jersey Cattle Society drew up a series of questions which were circulated amongst the members of the Society, whose answers are collated and form the basis of this publication.

The book is divided into chapters, the headings of which correspond in the main with the order of the questions. In many cases the answers were similar, and an analysis of such replies constitutes the ground work of the chapter, but answers in detail are also given to certain questions, where they have seemed particularly valuable. It must be understood that the responsibility for the accuracy of the information given in the replies rests with the contributors.

It has been impossible to quote from all, but none the less the answers sent in have been of the greatest assistance, and the Council of the Society take this opportunity of acknowledging the courtesy and assistance they have received from those breeders, who have kindly filled up and returned the papers.

---

[1] From information gained during a personal visit to these countries, Mr. Hall considers these figures, based upon various reports, somewhat too favourable. Both in Sweden and in Denmark many herds exist which doubtless produce milk of the quality stated; but the *average* throughout each country is much less. Neither at the Agricultural Laboratory, Copenhagen, nor at the Agricultural College, Upsala, do they admit a higher average production throughout each kingdom respectively than 1 lb. butter from 28½ lbs. milk.—ED.

QUESTIONS.

No.
1. GENERAL QUERIES—How long have you kept a herd of Jerseys?
2. What number of cows have you usually in milk?
3. Have you any experience regarding Island-bred Jerseys as compared with home-bred? As to their constitution? As to their dairy merit?
4. Do you select bulls for service on account of their personal appearance? Or for the dairy qualifications of their ancestry?
5. What is the acreage of your farm? How much arable? How much pasture?
6. What is its elevation—level above sea?
7. What is the general character of the situation?
8. What is the general character of the soil?

9. GENERAL MANAGEMENT OF COWS—What do you consider the best way of managing your cows in summer time?
10. Do you tether them?
11. Do you let them run loose in the field?
12. Do you advise Jersey cows being kept in entirely during the winter?
13. Do you give them exercise daily?
14. Do you prefer that they be turned out daily in the pastures for a few hours, weather permitting?
15. What months in the year do you take your cows under cover at night?
16. About what temperature do you keep your cow-sheds?
17. Do you give your cows chilled or cold water in winter?
18. Have they water constantly beside them?
19. Are they only given to drink at intervals?
20. Have they rock salt in the manger?

21. FEEDING—What do you consider the best kind of artificial food for cows from October 1 to April 30?
22. What quantity of hay and chaff do you give per day?
23. What kind and quantity of roots do you give per day?
24. Which foods do you consider the best for butter?
25. Which foods do you consider best for milk?
26. How many times a day do you think a cow should be fed? At what hours?
27. Do you milk your cows *before* or *after* giving them turnips or other food which has a strong odour?
28. Do you steam any or all the food, and by what system?
29. Do you mix hay, corn, and roots together?
30. How long before feeding do you mix them?
31. What artificial food do you find gives butter the *best* colour and texture in the winter months? What natural food?
32. What is the *worst* food for this purpose?
33. What concentrated food, if any, do you give your cows in summer when they are at grass?
34. What is your opinion of linseed cake as compared with cotton cake?
35. Or any other cake for the production of butter?

PREFACE

No.

86. CALVING—Do you think it better to milk a cow right up to the date of calving?
87. Or do you prefer that she should be dried off?
88. If so, how long before calving?
39. Do you think it best to milk out the cow entirely directly she calves?
40. Or do you prefer only to milk a little at a time and frequently?
41. What food do you give for the five weeks previous to calving in winter? What food in summer?
42. Do you keep the cows indoors for those five weeks?
43. Or are they turned out?
44. If kept indoors what exercise do they have daily?
45. What length of time after calving do you think should elapse before the cow is served?
46. At what age are your heifers served?

47. MILK FEVER—What treatment do you advise previous to calving to prevent milk fever?
48. How long before calving do you commence the treatment?
49. What medicines do you think best at that time?
50. If a cow shows signs of milk fever what treatment and medicines do you advise?
51. After calving how do you feed the cow for the first six days and when do you consider that all danger of milk fever is practically over?
52. Do you consider milk fever contagious?
53. Do you consider milk fever infectious?

54. ABORTION—Have you any experience of abortion or premature calving in your herd?
55. Do you think it is contagious?
56. What do you consider to be the principal causes of premature calving and abortion?
57. Do you know of any treatment likely to arrest the spread of abortion in a herd? Have you tried it? With what result?
58. Do you consider that abortion permanently damages the milking properties of cows?
59. How long after abortion or premature calving do you allow the cow to be served?
60. Do you believe a bull after service of a cow that has calved prematurely or has aborted, can infect other cows?

61. STERILITY—Do you find cattle that have aborted are apt to pass through a more or less protracted period of sterility?
62. How do you deal with this difficulty?
63. What proportion of cows do you think become permanently barren?
64. Have you ever had an outbreak of temporary sterility—that is, repeated turning from the bull's service in your herd?
65. Can you account for it?
66. What is your treatment for it?

No.
67. REARING CALVES—Do you take away the calf from the cow directly she has calved?
68. Or do you prefer leaving the calf with the cow for a few days?
69. What do you consider the best system of rearing calves?
70. What kind of food and what quantity per day do you give them from their birth until six months old?
71. What time and how often do you feed them?
72. What difference do you make in the rearing of bull calves and heifer calves?
72A. Have you ever had scour amongst calves? What is your treatment?

73. ENSILAGE—What is your opinion of silage?
74. Do you make it on the sour or sweet system?
75. What is the temperature of the rick when the weights are applied?
76. How does feeding with silage affect the quality and quantity of milk and butter?

77. BARRENERS—Do you fatten your "barreners"?
78. To what dead weight? State in lbs.
79. What price do you get for them?

80. DAIRYING—Do you sell milk or butter?
81. If the latter, what is your system of making the butter?
82. Do you separate the milk by means of a separator?
83. If so, what separator do you use?
84. Or if not, what system of cream raising do you adopt?
85. Do you churn sweet cream or ripened cream?
86. At what temperature do you churn sweet cream—In summer? In winter?
87. At what temperature do you churn ripened cream- In summer? In winter?
88. How do you dry your butter?

89. MILK RECORDS AND BUTTER TESTS—Do you keep milk records?
90. What is your system of taking them?
91. Do you take periodical butter tests of the general quality of the whole herd? Or of particular cows?
92. Have you any particulars of the results to offer?

93. DAIRY PROPERTIES—Do you believe butter properties to be transmitted by inheritance in particular families of Jerseys?
94. What do you consider to be the annual average weight of butter given by a Jersey cow? Under five years old? Five years old and over?

95. COST OF KEEP—What do you consider the total cost of keeping a Jersey cow for a year, excluding rent, rates and taxes and interest on capital, viz.: Cost of labour per cow? Cost of food per cow? All food to be charged at market price.
96. Have you any further suggestions to offer?

CONTENTS.

		PAGE
Chap. I.	Introduction	7
II.	General Management	9
III.	Feeding	14
IV.	Calving; Parturient Apoplexy, or Milk Fever	23
V.	Abortion; Premature Calving; Sterility	34
VI.	Calf Rearing	42
VII.	Cost of Keep; Dairy Properties; Testing Cattle; Barreners	46
VIII.	Dairying	51
IX.	Conclusion	54

CHAPTER I.

Introduction.

A TREATISE on the feeding and management of Jersey cattle would not be complete without a short history of the breed, but as this has been given at length in the first volume of the English Jersey Herd Book, a brief *résumé* will only be given here.

It would appear that the Jersey breed of cattle has been kept pure for over a century. In 1789 the States of Jersey passed a stringent law prohibiting the importation of cattle from France, and although cattle could then be imported from England, and from the other Channel Islands, there are not many instances of this being done. Indeed the only case of English cattle being imported for the purposes of breeding proved a failure, and it is believed the produce were all killed.

The cattle of the Island were noted for the peculiar richness of their butter, and the Jersey farmer, " conscious of possessing a " breed so excellent for the production of rich milk and cream, was " content to possess an ugly, ill-formed animal with flat sides, &c., " provided that she had a well-formed and capacious udder with large " swelling milk veins. The only question in the selection of a bull " amongst the most judicious Jersey farmers being ' Is the breed a " good one,'—meaning, had its progenitors been renowned for their " milking and creaming qualities."[1]

At that time the Jersey cow produced nearly twice the amount of butter yielded by the Normandy or Brittany cows, and that her prolonged period of lactation was recognised is shown by the statement that, with a small quantity of hay and roots, she produced a rich and well coloured sample of butter up to within six weeks of calving.

In the year 1834, the first attempt was made to improve the form and quality of the Jersey, and a scale of points for bulls, cows and heifers was drawn up. On that scale of points[2] prizes

---

[1] See the article on the Jersey—mis-named "Alderney"—Cow by Col. Le Couteur in vol. v. of the *Journal of the Royal Agricultural Society of England*, 1844.

[2] The Scale of Points is printed in the English Jersey Herd Book, vol. i., p. 23.

were awarded, and from that date the breed of Jersey cattle began to improve.

To give an idea of the butter properties of the Jersey cow, it may be mentioned that, in the year 1845, Mr. Hume, President of the Royal Jersey Society, tested three cows, two years old, for butter produce, and showed that they made annually a profit of over £15 apiece. Not being satisfied with this experiment, he made a further trial the following year, with the same cows, and showed an additional profit of about 33 per cent.[1]

In 1878, the English Jersey Cattle Society was established (subsequently incorporated in 1883), and in 1879 the first volume of the English Jersey Herd Book was published. As the number of members increased, and the entries at the shows became more numerous, a general improvement in the quality of the herds of the country took place, the breed becoming more popular as its valuable dairy properties were better recognised.

In 1886, butter test classes were started at shows, and in 1890, the English Jersey Cattle Society held a show at Kempton Park, Sunbury on Thames, where there were separate classes for English and Island bred cattle; also two butter test classes for old and young cows, every animal being weighed on entering the show. The food returns of the cows tested at that show were also given, but this has not been attempted since, as it has been found difficult to obtain accurate information. Butter test classes have from that time formed part of the leading shows in England, and have also been adopted on the Island of Jersey.

The above is a short outline of the history of the breed, and of the work of the Society. The history of the Jersey cow points a moral which cannot be overlooked, "Beauty and utility should be combined." Although always noted for her dairy properties, it was not until the show-ring points (which were indicative of good dairy cattle) were drawn up, and some approach to uniformity of aim arrived at among breeders, that the increased demand arose for Jerseys from other countries, with a consequent increase in their value. To sacrifice appearances even now to the dairy qualities, or the reverse, must in either case be wrong.

---

[1] See English Jersey Herd Book, vol. i., p. 35.

CHAPTER II.

General Management.

A STUDY of the names of breeders in the Herd Book will show that the Jersey is now kept in most parts of the kingdom. In many localities, especially in the northern counties, she is regarded as fit only for a gentleman's herd, although the excellence of her milk and butter is admitted. There is an idea among the farming community that she is too delicate to take her place on the farm, which is not correct, at any rate in the case of English-bred Jerseys.

As the geographical situation of the country is not a sure test of the climate, questions were asked as to the climatic conditions of the farms, their height above the sea level, the nature of the soil, and generally, as to the opinion of the constitution of English and Island-bred cattle.

It is known that Jerseys have been kept in England and Ireland for many years, and have adapted themselves to the climate. On the borders of Scotland at an elevation of 300 feet above the level of the sea, where a herd has been established nearly forty years, they are described as being "the best and hardiest of cows."[1]

A breeder in the Fylde district of Lancashire giving the result of his experience on his own farm as well as on an experimental farm of 160 acres where a mixed herd is kept, states that: "Home " bred Jerseys after three or four generations are as hardy as the " common cows of the country."[2] Another breeder in Buckinghamshire reports that, " The Island bred Jerseys require careful " treatment during the first winter and spring, but afterwards if " judiciously selected as to points indicating good constitution, they " are as robust as the average home bred animals."[3]

A breeder in Ireland, who has kept Jerseys for fifteen years near Dublin, considers the " Island bred animals to be somewhat

---

[1] Lady Marjoribanks, Tilmouth Park, Cornhill, Northumberland.
[2] Rev. L. C. Wood, Singleton Vicarage, Poulton le Fylde, Lancashire.
[3] Rev. S. H. Williams, Great Linford Rectory, Newport Pagnell, Bucks.

"delicate when they are first imported, but with care they soon "become acclimatised, and after a year or so there is little if any "difference between them. Perhaps, of the two, home-breds are "the most hardy, as they stand the climate remarkably well from "their birth."[1]

The farms on which they are located vary in elevation from the sea level to 600 feet above, and in character from the high and exposed limestones and cold clays to the lighter, more genial, and warmer soils. The animals, as a rule, are turned out in the field in the day time, and only taken in at night during the winter months from October to April, this being considered the most economical way of treating dairy cattle, since extra food must be given if animals are left out at night in the colder months.

In the winter, when the animals run out daily for a few hours, it is recommended that they should be taken in directly they stand about the gate. In most cases they are turned out the first week in May, provided the season is not backward, and they usually remain out, if the weather continues mild, until the end of September or even later.

In turning out cattle in the spring it must always be remembered that they will not, when they have once lain out, readily take to their winter rations again, consequently it is better to keep back freshly calved cows a little longer, and avoid having to bring them in again if the nights are inclement.

Where possible, the animals should have constant change of pasture, and should have different fields to run in for day and night. At some places in this country, and invariably in the Island of Jersey, the cows and heifers are tethered in the day time, being changed frequently to fresh ground. The system of tethering has much to recommend it where the grazing ground is limited, and the pasturage rich; there is little, if any, waste, a larger number of stock can be carried per acre, and if the field is a large one, and the season favourable for grass, what is not required for the herd can be made into hay.

Cattle, when turned out, should have their withers and backs dressed frequently with warble preventive dressings. Miss Ormerod's

---

[1] Mr. W. Milward-Jones, Rosebank, Rathfarnham, Dublin.

recipes are the most efficacious, and her recommendations should be carried out.[1] They are as follows:—

(a) Train oil rubbed along the spine, loins and ribs.

(b) or a mixture of 4 ozs. flower of sulphur, 1 gill spirits of tar, 1 quart train oil, mixed well together, and applied similarly once a week to the spine and loins.

(c) or a mixture of spirits of tar, linseed oil, sulphur, and carbolic acid, to be used in the same way.

Although it is a difficult matter to keep the cow houses at any special temperature, yet all cow keepers are agreed that dairy cows should be kept in sheds free from draught, but well ventilated. The walls should be lime-washed and disinfected twice a year at the least. Peat moss thrown down behind the cows after they are milked in the winter months helps to keep the places sweet, as it acts as a deodoriser, and prevents the valuable constituents in the manure from being wasted.

Water for cattle should always be pure and clean, and if slightly warmed in the winter will be found economical from a feeding point of view. Rock salt should always be in the manger.

The relative constitution of the English and Island-bred animal is a question on which there is difference of opinion. The majority of the English breeders regard home-bred Jerseys as stronger in constitution than those bred on the Island, while the minority consider that Island-bred cattle, if taken care of the first two winters, eventually become quite as hardy and as profitable as the English-bred animal.

It has been mentioned that Jersey cows were weighed at the Kempton Park Show. At that Exhibition, twenty-four English-bred animals averaged 5 years 8 months old, and 826 lbs. live weight, while thirty Island-bred averaged 5 years 4 months, and 735 lbs. live weight. These figures show that the Island-bred animals are about one-ninth less in weight than the home-bred ones.

If, however, the question of live weight is eliminated, there would seem to be little to choose between the dairy merit of the two classes of Jersey cattle. One of the aims of the Jersey breeder should be, to keep his animals near the normal weight of about 850 lbs. (the average of seven years' weighings at the London

---

[1] "Leaflet on Ox Warble-fly, or Bot-fly," by Miss Eleanor A. Ormerod, Torrington House, St. Albans.

Dairy Show), and to breed only from well authenticated dairy ancestors, disregarding entirely the question of the value of the carcase of the animal when done with. This, however, is not the universal opinion as regards the live weight of Jerseys, for some breeders think that there should be no limit to the size of the Jersey cow, contending that the larger animals are stronger in constitution, and yield more milk of equally rich quality; at the same time, some replies assert that with largeness of frame an increase of milk, though of poorer quality, is obtained, and that the smaller animals are the best for butter.

With regard to bulls, breeders agree that the sires in the herd should have good dairy ancestors for two or three generations, and with this, personal appearance should be considered. In the Island of Jersey a masculine-looking bull is preferred, while amongst some English breeders a bull of feminine appearance is often selected. In Jersey, where, in consequence of the warmer climate, and the different system of rearing, the cattle are naturally finer and consequently smaller than in England, it is thought best to use bulls of strong constitutional appearance, while in England, where they are apt to grow coarse, the use of smaller and more feminine-looking bulls is by some breeders considered advisable. At the London Dairy Show an attempt is made to secure both qualifications, viz., appearance and utility, by giving prizes by inspection to bulls descended from dams that have won Butter Test Prizes.

In most herds, it is the practice to get the heifers to calve at two years old, as the dairy qualities are thereby developed. When they calve earlier than this, it is better to let a longer period than usual elapse before they calve the second calf. The secret of the successful management of cattle generally, whether Jerseys or not, is to study the individuality of each animal. To put it shortly, the temperament and constitution of the animal must govern its treatment, and this supplies the key to successful management.

Notes on Management received from Breeders.

Mr. Richardson Carr, agent to Lord Rothschild at Tring Park, Hertfordshire, writes:—" If a herd of Jerseys is to be kept solely for dairy purposes regard-
" less of showing and appearance, I should turn the cows out during the day
" from about the middle of May if the weather is suitable. As the nights
" get warm I would also leave them out at night. If showing is considered,
" rather different treatment must be adopted, especially for the show animals,

"and if it is required that the whole herd should look in show condition,
"then I should not turn them out at night at all." Mr. Carr advises that
Jersey cows should be kept in entirely during the winter in order to get the
greatest dairy return. He would, however, give them exercise daily by
turning them out a few at a time into a sheltered yard, or in the case of
valuable cows would have them led out. The reason for this treatment is
that he does not think it economical to allow the temperature of a cow in
full profit to drop too low, there being very few days in winter when this
would not take place if cows are turned out in the field. He considers this
to apply to all dairy cattle irrespective of breed.

Mr. Crawford, agent to Mr. W. O. Hammond, St. Albans Court, Wingham, Kent, turns out all the animals daily and considers it a "matter of
"great importance if you desire a healthy herd. Probably in cold weather
"the quality and quantity of milk may suffer to some extent, but it is more
"than counterbalanced by the increase that takes place by being in a hardy
"good state of condition during the year."

Mr. Edward Colston, of Roundway Park, Devizes, Wiltshire, whose land
lies 400 feet above the sea, remarks that he has always been struck with the
extraordinary hardiness of Jersey heifers. From about a year old they are
turned out in the Park, where they have absolutely no shelter, summer or
winter, no matter what the weather is, until within two or three weeks of
calving. They run with the deer and take their chance of food with them.

Mr. Herbert Padwick, of Manor House, West Thorney, Emsworth, who
has kept Jerseys for twenty years, and usually has from 80 to 120 cows in
milk, says:—"As to bulls, I have always found that imported animals,
"or those bred in England from imported sires and dams, get stock very
"much hardier and with better constitutions than those bred from a line of
"English ancestors; this has surprised me, but the fact is incontrovertible.
"Their stock is also much more docile, a very important point. When the
"young heifers have a very wide range, the wildness of the English bred ones
"causes them to cast their young. As to cows, what few imported ones I
"have had I have found very delicate, and unless tenderly nursed and highly
"fed they are unequal to the home bred ones. I am very strongly of opinion
"that all animals should be bred upon the farm where they are to live."

Mr. W. Buckley Roderick, Fronheulog, Llanelly, Carmarthenshire, turns
out his heifer calves that are born in the winter or early spring as soon as
the weather is warm. They remain out in the fields day and night, winter
and summer, in all weathers until they are about to calve. For the first
year, heifers have an open shed to run into, but the yearlings have no
shelter of any kind. They appeared to be thoroughly healthy, and Mr. Roderick has never lost a single animal by this treatment. They look better in
the early summer and pick up quicker than if they had been housed all the
winter.

Colonel Willan, of Thornhill Park, Bitterne, Hants, is of opinion that
"the best management for cows in the summer is to give them the best
"pasture one can, and change them from one field to another as soon as *they*
"*appear* to get stale. To form that judgment it will be found most useful
"to weigh all the milk, as the period when a pasture should be changed can
"be determined by the falling off in milk. It is wonderful the stimulus to
"milk yielding, which a change from one pasture to a fresh one often proves
"to be."

CHAPTER III.

Feeding.

The feeding of Jersey cattle differs much from the practice prevailing in the feeding of other breeds, for the reason that if a Jersey is to be profitable from an economical point of view she must not be over-fed, but receive just so much food as will keep up her milk to the proper standard, in quantity and quality. The best and most natural food for Jerseys is the early grass in the spring of the year, which generally lasts up to the second week in June. The aim of the careful feeder should be to try and get a food as near as possible to the composition and quality of good grass.

With such rotation crops as rye, trifolium, lucerne, sainfoin, and maize (where it can be grown), the flow of milk can be kept up when the grass begins to get old, but when these crops have in their turn got coarse and woody, the root crops must be looked to as substitutes. Here the difficulty of feeding commences. Where milk for sale only is required there is not much trouble, but, when butter is to be made, great care is necessary. The same remarks hold good with the concentrated foods, under which term are included cereals and the various meals and cakes. For milk production cabbages, carrots, swedes, mangels, silage, grains both fresh and dried, crushed oats, bran, cotton and linseed cakes, and some of the manufactured compound cakes are all recommended; but where the milk is to be used for making butter they are not all equally suitable.

With careful feeding and skilful manufacture, the butter should be uniform in colour and appearance. In winter, butter is naturally a trifle paler in colour, especially where the only available roots are mangels, and to counteract this the artificial foods which improve the colour of butter should be used. With respect to roots, most feeders are agreed that swedes and turnips are the worst possible food for butter, and brewers' grains come into the same category. Carrots, parsnips, cabbage, kohl rabi, kale, mangels, potatoes and sweet silage, are all good for butter, and of these carrots, parsnips, cabbage and kohl rabi are undoubtedly the best. Where carrots and parsnips can be grown there is really no better food, as they will keep good throughout the whole of the winter, but as they will not do in all soils, cabbage, kohl rabi, and mangels must take their place.

In feeding with cabbage, the stalks and outside leaves must not be used, and the hearts or middles are better when pulped, and mixed with chaff. As frost gives cabbage a strong flavour, it is not safe to feed with it after the winter has set in, and therefore the crop should be grown for use in the autumn and early winter months. Where cabbage is fed alone it is preferable if it can be given out of doors. The best kinds are the Enfield Market, followed by Drumhead, Thousand Headed Kale and Drumhead Savoy. Potatoes, if plentiful, can be used after the cabbage is done and before the mangel is ready, but they are better steamed.

Sweet silage does not appear to be generally adopted, although in the few cases where it has been tried it is said to impart a good colour and flavour to the butter, and to be a good substitute for grass, but it should be given after milking is over, so that the milk does not get tainted with the smell. It is probably for this reason that it is not more generally used.

All roots, including kohl rabi, kale, cabbage, &c., should be pulped and mixed with chaff. The chaff may consist of equal quantities of hay and sweet oat straw, and the mixture should lie twenty-four hours to ferment. Common salt is often put into this mixture. Some recommend steaming the food, while others consider that hot water thrown over the heap is quite as efficacious. Concentrated food should be given with chaff, and where it can be evenly distributed, it is a good plan to mix it with the chaff beforehand; but as it is difficult where there are a large number of animals to ensure that each cow gets its proper ration, it is perhaps advisable to give each cow its separate allowance of concentrated food at each feed. Some think that meals should always be mixed with hot water, and allowed to soak for some little time.

Of the concentrated foods, crushed oats, bran, bean, and maize meals seem to be generally approved; but there is a remarkable difference of opinion as to the relative values of linseed and cotton cakes. Linseed cake is thought by many to make the butter strong, soft, and greasy, and no doubt if given in large quantities this might be the case. Cotton cake, on the other hand, is said to make butter of good texture, but to affect the health of the animals and to be dangerous to in-calf cows if given in large quantities. In most of the feeding rations mentioned in the replies too much cake is given, and it is believed that if cake were used in moderate quantities, its deleterious effects in butter would disappear.

In Fleischmann's Book of the Dairy it is stated that 2¼ lbs. of cake is the outside quantity for a milking cow, and from the article on the production of milk rich in fat, by Mr. N. H. J. Miller, Ph.D., in the eighth volume (part 4) of the *Royal Agricultural Society's Journal* for 1897, where the whole subject is most carefully considered, it would seem that 2 lbs. to 3 lbs. of cake is as much as any small cow fed for butter produce should have per day, in addition to the other artificial foods.

The quantity of food that should be given to a Jersey in full milk has not been very clearly defined in the answers sent in, but the general opinion seems to be that from 6 lbs. to 8 lbs. of artificial food, 10 lbs. to 15 lbs. of hay and chaff, and 10 to 12 lbs. of roots is the proper winter allowance for a cow of about 850 lbs. live weight. In the case of very heavy milkers the quantities, with the exception of cake, may be increased proportionately, as it must be remembered that the dairy value of the cow is dependent upon her capacity for assimilating a large quantity of food and producing butter fat; other things being equal, her value depends upon this power. Of the hay about 5 lbs. to 8 lbs. may be given long. It is a difficult question to say of what the artificial food should be compounded. What suits one cow will not always be palatable to another, and a change of food is almost as necessary to a cow as it is to a human being, although the changes should be very gradual.

To get at the most suitable diet for the production of butter, it will be well to consider the different sorts of food recommended, and the influence they have on the butter produced. Cabbage, kohl rabi, kale, and carrots, all produce butter of good colour; mangels on the other hand, make a very pale butter; linseed and decorticated cotton cake, maize and bean meal, produce a deeper coloured butter than do crushed oats, and bran. It would seem, therefore, that where cabbage, carrots, and other roots, are being given, crushed oats and bran should be the staple food, while, after Christmas in those parts of the country where mangels are the only available roots, cakes and maize and bean meal are better than oats and bran.

The food ration, as applied to cattle, defines the quantity of certain constituents which must be present in the food, in order not only to sustain a vigorous life, but also to secure the best economic results in fattening, or in milking.

These constituents are—dry organic matter, flesh-formers or albuminoids, heat-formers or carbohydrates, and fat.

In the early days of chemistry and physiology, it was supposed that the fat alone in the food supplied the material from which fat was transferred to the body or conveyed to the milk. At a later stage this theory became untenable, as it was shown that the amount of fat present in the food was quite insufficient for the purpose. It was next asserted that the albuminoids or nitrogenous matter supplied the deficiency. Subsequently, however, doubts were raised whether fat, plus albuminoids, was sufficient to account for the production of animal fat. Recently it has been discovered that the carbohydrates in the food ration contribute largely to the formation of butter fat, this conclusion being supported by a careful series of experiments recently conducted at Geneva, New York.[1]

The food ration may be varied to an almost indefinite extent, and still be economical and efficient, provided that the proportions of the different chemical constituents be strictly regulated; but when this precaution is neglected it is frequently found to be the reverse of economical. For instance, an excessive use of cake only produces the result of inflating the bills of the farm, and enriching the manure heap; whereas a deficiency of albuminoids reduces the condition of the animal below the standard of vigour, and renders it incapable of yielding profit.

---

[1] See the article on "The Source of Milk Fat," by Professor R. Warrington, in the *Royal Agricultural Society of England's Journal*, 3rd series, vol. ix., p. 2, June, 1898.

We subjoin four examples of feeding rations.

<table>
<tr><td colspan="2">EXAMPLE I.</td><td colspan="2">EXAMPLE II.</td></tr>
<tr><td>Carrots</td><td>12 lbs.</td><td>Drumhead cabbage, inner leaves</td><td>12 lbs.</td></tr>
<tr><td>Chaff, oat straw</td><td>5 ,,</td><td>Chaff, as in Example I.</td><td>10 ,,</td></tr>
<tr><td>,, good hay</td><td>5 ,,</td><td>Linseed cake</td><td>2 ,,</td></tr>
<tr><td>Decorticated cotton cake</td><td>2 ,,</td><td>Crushed oats</td><td>2 ,,</td></tr>
<tr><td>Crushed oats</td><td>2 ,,</td><td>Bran</td><td>2 ,,</td></tr>
<tr><td>Coarse wheat bran</td><td>2 ,,</td><td>Good meadow hay</td><td>7 ,,</td></tr>
<tr><td>Hay, good</td><td>7 ,,</td><td></td><td></td></tr>
<tr><td colspan="2">EXAMPLE III.</td><td colspan="2">EXAMPLE IV.</td></tr>
<tr><td>Mangels</td><td>14 lbs.</td><td>Mangels</td><td>14 lbs.</td></tr>
<tr><td>Chaff, as above</td><td>10 ,,</td><td>Chaff, as before</td><td>10 ,,</td></tr>
<tr><td>Decorticated cotton cake</td><td>3 ,,</td><td>Decorticated cotton cake</td><td>2 ,,</td></tr>
<tr><td>Maize meal</td><td>3 ,,</td><td>Maize meal</td><td>2 ,,</td></tr>
<tr><td>Hay, good</td><td>7 ,,</td><td>Malt, sprouted</td><td>2 ,,</td></tr>
<tr><td></td><td></td><td>Hay, good</td><td>7 ,,</td></tr>
</table>

The times at which animals should be fed must be regulated by the system of management adopted in the particular herd. Where the cows are turned out the number of feeds is fewer than where they are kept in. In the latter case, the quantity of concentrated food is not necessarily increased, a little more hay or straw being usually given. In the autumn months, about 2 lbs. of concentrated food of an astringent nature may be given with advantage, undecorticated cotton cake and crushed oats being those generally used.

Bulls should be kept in good hard condition. The addition of crushed oats and a little linseed cake to the ordinary ration of roots and chaff is perhaps the best food. In summer, they should have green food, lucerne and cabbage, or a little rye and cut grass for choice. Vetches are not recommended. It is most essential that they should have plenty of regular daily exercise, if possible, on a road.

If a Jersey heifer calf is fed for show, and kept too long on milk and other rich foods, she is apt to grow fat and to develop a large fleshy udder, which, while for the time it may promise well, eventually turns out deceptive. The natural growth and expansion of the stomach in the young animal should be encouraged and assisted by natural feeding, but it is unwise to give large quantities of highly concentrated food to yearlings.

The best rations for heifers are as follows:—In the winter months, pulped swedes mixed with chaff—half oat straw and hay—

the quantity to be regulated according to the age and size of the animal; in the summer months, they should run in a grass meadow separate from the cows. In-calf heifers should not be starved, as abortion is sometimes attributed thereto. Rock salt should always be within their reach, and they should have their backs dressed to prevent the attacks of the warble-fly. If artificial foods are required, small quantities of linseed cake, oats, and bran are recommended.

Notes on Feeding received from Breeders.

In Mr. John Tremayne's herd at Heligan, St. Austell, Cornwall, where Jerseys have been kept for over forty-five years, the cows flush in milk are fed twice daily with the following mixture: 3 quarts bran, 3 quarts bruised dredged corn, 1 pint ground decorticated cotton cake, and ½ pint maize meal, 8 lbs. of hay and 8 lbs. of straw per day, and 28 lbs. of mangels and cabbage. The roots are given after the cows have been milked. The food is not steamed. Mr. Tremayne considers linseed cake the best for young cattle and decorticated cake the best for milch cows.

Mr. Herbert Padwick, who feeds from 80 to 120 Jersey cows, thinks oat meal, malt dust, decorticated cotton cake, linseed cake, and maize, the best kinds of artificial food for cattle. He gives 8 lbs. of hay and 4 lbs. of oat or wheat straw chaffed, and as much long straw as the cattle will eat per day, with 30 to 40 lbs. of mangels and no other roots. He tries to mix the foods as nearly as possible to Mr. Lloyd's ratio, but allows rather more fat-producing foods—bean meal, oat meal, decorticated cotton cake, if the butter buyers will allow its use (which is not often the case, since it affects the flavour of the butter), linseed, and linseed cake, in the order named. His cows are fed at 4.30 a.m., 3 p.m., and with straw only at 4.15 p.m., but if kept in by stress of weather they have an extra feed at 9 a.m. The chaff, corn, and roots are mixed together from six to eighteen hours before using, but the cake is given separately, so that the heaviest milkers and animals which are not doing well may get extra food. Mr. Padwick considers bean meal, bran, and malt dust give butter the best colour and texture in the winter months, but he has generally noticed that the firmness of the butter is most marked when a good mixture is used, that is to say, roots, hay, straw, and concentrated food. He is of opinion that linseed cake can always be used in moderation without affecting the flavour of the butter—cotton cake never. The worst food for the colour of butter is mangel, and the worst foods for the texture are linseed and oil cakes in excess. In summer the foods used are a little bean meal, oatmeal, and malt dust. Over a series of years in his herd 7 quarts of milk produced 1 lb. of butter.

The Hon. Alex. E. Parker, agent to Earl Cadogan, at Culford, Bury St. Edmunds, Suffolk, writes:—"When turned out in summer the only con-"centrated food the cattle receive consists of from 2 lbs. to 3 lbs. of linseed "cake, bran and crushed oats, which is given at milking time. In the

"winter when in full milk the cattle receive either 4 lbs. of bran, 2 lbs.
"of crushed oats and 1 lb. of linseed cake, or 3 lbs. of bran, 2 lbs. of crushed
"oats and 2 lbs. of linseed cake, with about 10 lbs. of long hay, 4 lbs. of
"chaff, carrots in winter and a few mangels in the spring. The chaff,
"corn and roots are all mixed together and steamed. The mixed food is
"given twice a day, morning and afternoon, and the hay at other times."

Mr. W. Adams, who has kept a herd near Gloucester for fifteen years, considers that, with the care due to all animals of dairy merit, the Island-bred animal has an equally good constitution if it has been healthily reared. As regards the production of milk and butter, his experience has on the whole been in favour of the Island-bred animal, as the increased size of English-bred Jerseys is an economic loss with no equivalent corresponding advantages, his motto being the maximum amount of butter at the minimum cost. He gives 6 to 8 lbs. of artificial food composed of Foster's dairy cake, crushed oats, germ sharps containing 35 per cent. of germ, with decorticated cotton meal and sometimes maize meal. Carrots, parsnips and mangels are the only roots used, and they are mixed with the artificial foods and chaff. As a change he found one-third palm-nut meal, one-third cocoanut meal, one-third rice meal, with a little decorticated cotton cake to be very good rations. All the food is mixed in bulk and boiling water poured over it, leaving it for 24 hours. Mr. Adams considers that whatever food he gave he could not dispense with decorticated cotton cake.

In Mr. C. C. Tudway's herd at Wells, Somersetshire, the allowance is as follows:—A liberal quantity of hay and chaff per day, 10 lbs. of carrots or parsnips, 4 lbs. crushed oats, 1 lb. bran, 1 lb. Indian meal and 1 lb. bean meal. Boiling water is thrown over the prepared food, which is left to ferment a few hours before using. The cattle are fed twice a day, at 6 a.m. and 4 p.m., and have a handful of hay given them the last thing at night. This allowance of food is given in the winter to animals in milk; in the summer they are out on the pastures day and night. Mr. Tudway considers that linseed cake is the very worst food for butter production.

Miss Standish, late of New Park, Brockenhurst, Hants, and now of Marwell Manor, Eastleigh, who has kept Jerseys for about sixteen years and has been successful in the butter test competitions, has experienced no trouble as a rule with imported stock, but thinks care is advisable during the first winter. She gives rather a full ration of artificial food. Cows in full milk get 6 lbs. of decorticated cake, 2 lbs. of crushed oats, 1½ lbs. of bran. Maize meal sometimes takes the place of grain, and "One and All" dairy cake the place of the cotton cake. Cows stale in milk receive 4 lbs. of cake, 1 lb. of bran and 1 lb. of crushed oats. Eight to 9 lbs. of hay and chaff and about 15 lbs. of cabbage is the allowance for cows in full milk, while 6 to 7 lbs. of hay and chaff and 8 to 10 lbs. of cabbage or mangel is given to those getting stale. In her opinion decorticated cotton cake, crushed oats and bran, combined with cabbage, make butter of the best colour. Her cows are fed twice a day, unless the weather prevents them going out, when the rations are divided into four. The cake, oats, bran and meal are mixed with warm water and stand for twelve hours before use.

The Rev. S. H. Williams, of Great Linford Rectory, Newport Pagnell, gives 2 to 3 lbs. of undecorticated cotton cake to cows in full milk in the summer when they are at grass. In the winter cotton cake, crushed oats or crushed maize and bran and a small quantity of grated mangels—about 6 to 7 lbs.—are mixed with the chaff, or in place of the roots a little boiled linseed or scalded linseed cake is poured over the mixture. The cattle are fed four times a day. In his experience maize meal gives a better colour to butter than crushed oats, while bran certainly tends to make butter pale. He prefers cotton cake to linseed cake.

One of the most successful breeders of butter cattle considers that cotton cake is the most powerful stimulant, but one of the most injurious foods—garget is common after large quantities, and so is death. The unrestricted use of it should not be given to the cowmen, or loss of good cows will be the result.

His feeding rations when there is little or no grass are as follows:—

5 lbs. dried distillers' grains
3 lbs. crushed oats   or   4 lbs. dried distillers' grains.
3 lbs. crushed wheat   4 lbs. crushed oats.
                                  2 lbs. crushed wheat.

The wheat must be crushed and soaked for twenty-four hours previously. The grains and oats for twelve hours.

15 lbs. of good hay chaff, to be increased to 20 lbs. if no cabbage is available.

Cabbages to be given in the fields, from October to February or even March, 15 lbs. per head, the outside leaves to be stripped off and the cabbage cut into four pieces. They should be given three hours before milking. From March to April about 12 lbs. parsnips are given.

All the cattle in milk are bedded up with good oat straw, of which they eat a fair quantity.

Heifers with first or second calf receive about 2 lbs. of linseed cake per day, in addition to their other food.

Mr. W. G. M. Townley, Hard Cragg, Grange-over-Sands, Lancashire, considers palm nut meal most valuable as a food to produce the nutty flavour most prized in butter.

Mr. J. F. Hall, Sharcombe, is of opinion that decorticated cotton cake and maize meal are the best kinds of artificial food for cows from October 1st to April 30th. For the average Jersey cow 15 to 20 lbs. of hay and chaff and 12 to 15 lbs. of carrots or parsnips is the allowance in his herd, besides cake and meal. He considers the following the best foods for butter in winter:—decorticated cotton cake, maize or bean meal, carrots, sound meadow hay, sweet oat straw and sweet silage of grass or green maize. For summer, grass with decorticated cotton cake if necessary. The cows are fed four times; at morning milking, noon, afternoon milking and night. In Mr. Hall's opinion linseed cake makes a soft greasy butter, and most cakes other than decorticated cotton cake are more fitted for producing flesh and fat. Mr. Hall sets a high value upon silage which he makes on the sweet system. He considers that sweet silage increases the quantity of the milk, and imparts a better colour to the butter.

Col. Willan considers sweet silage, that is, silage which is heated to at least 140°, most valuable food if made from good grass, clover, trifolium, rye, vetches, &c., but that it is a mistake to think anything will make silage, any more than anything will make hay. The silage on his farm is made on the sweet system, and he thinks that feeding with silage improves the quality of the milk and the colour of the butter—the weighing of the milk will show this. In his opinion silage will not make butter taste unless it is left in the cow-house to infect the milk with its smell—certainly not through the stomach of the cow.

Mr. Richardson Carr, agent to Lord Rothschild, writes:—"I find "the best concentrated foods for cows in the winter are ground oats, bran, "linseed cake, bean meal, and a little wheat meal; the quantity of long hay "per day that should be given to milking cows is 6 lbs., with 10 lbs. chaff, "half oat straw and half hay, 10 to 12 lbs. of carrots, parsnips or cabbage "early in the season, and 12 lbs. of pulped mangel late in the season make "up the rations. If kept in, the cows should be fed four times a day."

Mrs. Thackwell, Rostellan Castle, County Cork, states that they make silage on the sweet system when the season is not fit for making hay, as it is believed to increase the milk and to give the butter a good colour. Sour silage, on the contrary, causes the butter to taste strong.

Mr. D. D. Crawford, agent to Mr. W. O. Hammond, St. Albans Court, Wingham, Kent, thinks that no dairy should be without silage; it is absolutely necessary where butter of high quality is desired; it increases the quantity of milk, greatly improves the quality and texture of butter, and is a capital substitute for roots of all kinds without giving the risk of a bad flavour and odour.

Mr. J. T. Keddie, agent to Sir James Blyth, Bart., Blythwood, Stansted, Essex, writes:—"The odour of silage is very objectionable and is taken "up by the milk if in the same building in which the cows are milked, but "it is good for dairy cows if mixed with hay and given to them out in the "fields when the grass is scarce." He considers the sweet system is preferable for dairy cows, but thinks that although it invariably tends to increase the quantity of milk, anyone with a good sense of taste can detect it in the butter.

CHAPTER IV.

Calving; Parturient Apoplexy, commonly called Milk Fever.

It is generally conceded, that there is always more risk attending the calving of dairy cattle, than of those which are kept for their beef-producing qualities. No exhaustive treatise on milk fever has hitherto been published, although most of the excellent books on cattle pathology prescribe various remedies for the disease. As these books are generally written by members of the veterinary profession who necessarily have not had the experience of the breeder, they do not go sufficiently into the treatment of cows previous to calving, which, from the opinions quoted below, appears to be a most important factor in the prevention of the disease.

All breeders seem agreed on one point, viz., that parturient apoplexy (called for brevity "milk fever" in this chapter) is caused by allowing the animals to get into too high condition before calving, and those who have been the most successful in averting the disease attribute it entirely to the preventive measures which they have adopted. Milk fever rarely, if ever, attacks an animal before it produces its third calf, and therefore, there is little, if any, risk with heifers or cows calving their second calf. The general course of treatment of cows due to calve is as follows:—

The cows are usually dried off from four to six weeks before calving, as they milk better after a few weeks' rest, although one breeder suggests that in particular instances it is wise to milk a cow right up to her time of calving, as a precaution against milk fever. About four to six weeks before calving they should be taken up from the field and fed sparingly. They should be exercised daily, the object being to get them into a natural state of health, and to do away with any superabundance of internal fat.

Some breeders recommend putting animals due to calve on a bare pasture for a short time previous to calving. This is probably a mistake: for a bare pasture generally contains fattening and nutritious food. To illustrate this, in dry summers stock always appear to thrive better than when there is an abundance of grass, especially if they have water and shade. On this account it is not prudent to put cows on a bare pasture before calving, it being better to take them into the yard, and feed them on dry food. Two or three days after this they should have an aperient drink, and from that time up to the date of calving, their bowels should be kept open, if possible, by judicious feeding, and they should be exercised daily. It is considered by some a mistake to give a drench just before calving, as after the physic has worked a reaction sets in, and temporary constipation follows at the very time when it is most harmful. Cows should be taken into the calving box and littered on peat moss for a week before calving; if straw is used for litter the cow should be muzzled to prevent her eating it, as it is always best for a cow to calve with an empty stomach.

The use of the clinical thermometer is recommended for a fortnight previous to calving, and if the temperature of the animal remains normal ($101\frac{2}{5}°$ to $102\frac{2}{5}°$ F.), there need be little cause for anxiety. Should the temperature of the animal go down (and there is considerable evidence that in milk fever this is usually the case—the thermometer sometimes reading as low as 95° F.) she should be given a little linseed cake, or any easily digested food of a laxative nature (care being always taken that the bowels are kept open); it will then be found that the temperature will return to its normal state. If the temperature should rise, and symptoms of milk fever are not present, other complications may be expected.

In most herds preventive medicines are used. Linseed oil and Epsom salts are the well-known remedies, but the extracted replies from different breeders give sufficient information on this subject to enable the reader to use his own judgment.

As soon as the calf is born, if left with the mother, it should be rubbed over with a little salt. The cow should be covered with a rug, if the weather is cold, and kept warm and free from draughts. Care must be taken that the cow cleanses properly, and in the event of that not taking place, the vulva should be sponged with disinfectants, and injections administered. If this treatment is not successful, a veterinary surgeon should be called in.

The cow should be milked frequently, but only a little milk should be drawn off at a time. For the first three days after calving a cow should not be left for long, and she should be visited frequently in the day and at night, as a change may come over her at any moment. This a careful cowman will at once notice, and in all probability, by prompt treatment in the early stages of the malady, will save the animal's life.

The temperature of the cow should be taken every two or three hours, and if it remains normal, it may be taken for granted that the animal is going on well. Assuming this to be the case, the food for the first three days should consist of hay tea, *i.e*, hay scalded in a bucket of hot water and allowed to stand until cold, and a few bran mashes. The scalded hay as well as the tea should be given as it counteracts the ill effects of too many bran mashes, which are apt to clog the stomach and interfere with the digestive organs.

Although all danger from milk fever is said to be over after the third day, great care should be taken for the first ten days that the animal does not catch cold, and that she is not over-fed. She should be brought to her food by degrees, and only such food as is easily digested should be given her at first.

The treatment of heifers differs from that of cows in one respect, viz., that there is not the need of preventive treatment, as there is little, if any, risk of parturient apoplexy with the first calf, and even with the second calf the cases are rare, and usually very mild in their attack.

A drink should be given after calving, and the animal should be kept warm. It is generally better to leave the calf with a heifer for ten days or a fortnight, until the udder gets perfectly right. Great care is necessary with heavy milking heifers, and they should on no account be turned out to grass until all appearance of inflammation in the udder has disappeared.

The symptoms of milk fever are described in most books of cattle veterinary practice, but in scarcely any of them is the low temperature of the animal commented on. Those who have unfortunately had experience of this disease know that there is no fever, but that, on the contrary, the temperature goes down.

As indicated above, the thermometer is the best guide with calving cows, and it may be taken for granted that with normal temperature there is not much danger. If the clinical thermometer shows a low reading, the animal should be treated at once for parturient apoplexy, and the remedies decided upon should be used. The choice of these must be left to the breeder, but it would appear that stimulants such as whiskey are the most successful, though homœopathic treatment has many followers.

If the animal drops, she must be kept well propped up, by using small bundles of straw, or sacks filled with straw, as they do not slip away like large or ordinary sized trusses of straw, and consequently make it much easier to keep the cow in an upright position. Her spine should be blistered with mustard or some equally strong liniment, and cold water or ice applied to the head. The udder should be stripped occasionally, and, if necessary, the catheter and the injection pump must be used. Her legs should be bandaged in order to keep the extremities warm, and she should be moved from side to side every two or three hours. When she loses consciousness, it is dangerous to make her swallow, and for this reason medicines in concentrated form are preferable. Homœopathic treatment is practised in the herd of Mr. J. R. Corbett and many other breeders, the late Mr. George Simpson being one of the first to introduce it. If consciousness is regained, the cow will recover, and it is surprising how soon she will get all right; but great care must be taken in feeding, and if she has a relapse the chances of saving her are reduced.

During the present year (1898) a new cure for milk fever has been brought forward, and from the reports which have been received it appears to have been successful.

The treatment consists simply in injecting iodide of potassium into the udder through the teats.

The udder and teats should first be carefully washed with water and soap (in which a little carbolic acid has been mixed), and then thoroughly dried.

The following mixture should then be injected, care being taken that the injecting syphon has been disinfected in carbolic acid and water.

One drachm of iodide of potassium mixed with one quart of boiling water, to be cooled down to 98° F. before being used. Half a pint to be injected into each quarter. A proper injecting teat syphon should be used, this can be obtained from any Veterinary Instrument maker.

The animal must then be placed in a comfortable position on her chest, and the udder may be rubbed gently.

The position of the cow should be changed every two or three hours.

Only one injection is necessary.

If milk fever is in the nature of apoplexy it will almost certainly be non-contagious. There are some breeders, however, who consider that it is infectious. Whether right or wrong, it is in any case extremely essential that the calving boxes should be kept clean and frequently disinfected.

The best way to disinfect a cowshed is as follows:—

Wash the floor and the lower parts of the walls with a solution of carbolic acid or creosote, and lime-wash the upper parts of the walls; then close all doors and windows, stop up any crevices with straw or hay, and burn 6 lbs. of sulphur in an iron pot (using two or three pots if the size of the cowhouse warrants it) until the air is so dense that you cannot see across the house (*i.e.*, by looking through the window from outside). Leave the cowhouse shut up in this state for at least twelve hours.

The rugs and utensils should also be kept clean and disinfected.

Cows are sometimes affected with "drop," or more correctly "*adynamia nervosa generalis*." This attacks animals both before and after calving. There is no fever, but the cow appears unable to rise. Warmth, good feeding, frequent milkings and doses of nux vomica usually effect a cure, but animals are generally much pulled down, and require attention afterwards. Whether or not this is another form of milk fever is open to doubt.

Notes on Calving and Milk Fever received from Breeders.

In reply to the questions on milk fever the following interesting account was received from Dr. Alfred Brown, of Summerlands, Broad Oak, Heathfield, Sussex, which he has kindly placed at the disposal of the Society.

Parturient Apoplexy, commonly known as "Milk Fever," its Nature and Mode of Prevention.

For several years previous to 1862 I had lost so many Jersey cows from the affection, then known as milk fever, that I determined to investigate the disease, and if possible discover a mode of treating parturient cows that would prevent its occurrence. Accordingly, when I next lost a cow from milk fever I made a very careful *post-mortem* examination of the fresh dead cow. This I managed very easily with the assistance of the knacker's man, who was very dexterous in the use of the knife and saw. The skin and horns were first removed, then the thoracic and abdominal viscera, all of which I examined carefully, also the udder. All the limbs were then taken off, and the ribs taken off near their origin at the spinal column. The ossa innominata were also removed, leaving only the head and spinal column. These were very carefully opened, the latter with a chain saw, so as to expose the brain and spinal cord. I carefully opened the covering membranes of both, which were very much congested, and there was a considerable amount of serum in both, and in places patches of effused blood. The kidneys were also congested, and there was very little urine in the bladder. I had examined the urine before the death of the animal, and found it loaded with albumen.

From these appearances I came to the conclusion that the cause of death was serous apoplexy. Serum is the watery part of the blood, and contains a large quantity of albumen. The symptoms during life, after the animal is attacked, are first, a heavy dull appearance of the eyes, and when recumbent inability to rise, and for this reason the disease is sometimes called "drop;" the temperature is low and the pulse slow. There is a total absence of all the symptoms of fever. Coma and death end the case in from one to three days. For about a fortnight before a cow calves there is a great determination of blood to the udder, in order that there may be a secretion of milk for the coming calf, which, while in utero, is nourished directly from the blood of the mother, through the placenta which is attached to the inside of the uterus, and communicates with the calf by the umbilical cord.

After the birth of the calf the cord is sundered by the action of the cow rising, and the placenta and membranes, which latter contained the calf in the liquor amnii, are expelled by the contraction of the uterus, after which contraction there is no longer any large quantity of blood sent to the uterus, but a much greater quantity is sent to the udder, and if the cow be in too good condition is apt to cause congestion of that organ, and the pressure of blood in the vessels prevents the proper secretion of milk, which would otherwise be secreted and removed by the calf. Not only is the udder congested, but the kidneys also, and the secretion of urine is suppressed, which greatly complicates the case.

Jersey and Shorthorn cows are more subject to this affection (parturient apoplexy) than any other breeds, but I have seen cases in Norfolk polled, Galloway and Ayrshire cows. Cows in a natural state would probably not be attacked with this affection, as they would not develop large udders. Breeding cattle for the production of milk and butter, and allowing the udder to be distended with milk, is of course unnatural, because in a state of nature the calf would run with the mother and would never permit the udder to be distended. However, the requirements of the dairy necessitate that cows should be treated unnaturally, and hence their liability to this affection.

I think I have now shown clearly the nature and course of the disease called milk fever, but which is really parturient apoplexy. In some instances when cows drop, there is only congestive apoplexy, that is, no effusion of serum, or blood, on the brain or spinal cord; such cases with proper treatment recover, and the cows, with good management before their next calving, are none the worse.

This disease, parturient apoplexy, is caused by cows being kept in too good condition before calving, and the treatment should be prophylactic, that is, preventable; this can be done by careful feeding before the cows calve, gradually reducing the food, so that they may be getting out of condition when they calve. I have found the best plan in winter is to keep the cow by herself in a loose box with a yard attached, giving a limited supply of hay, and if necessary an occasional mash, of two quarts of bran with a pint of linseed oil mixed with hot water—all cows will eat it readily; and after calving, giving no hay for the first two days, only hot bran mashes, then giving hay in moderation, gradually increasing it, and one gallon of bran wet with hot water three times a day. At the end of the sixth day, the cow milking pretty freely, and the udder softening, she may have a full allowance of hay and three bran mashes a day, each containing one gallon of bran, and a pint of decorticated cotton cake meal, which will improve her milking and condition. In summer the cows should be taken up from grass and placed separately in loose boxes, and given mown grass in moderate quantity, not clover, a little hay and if required a bran mash with a pint of linseed oil in it occasionally, made the same as I recommend for winter. Deep milking cows have the most severe attacks, and consequently require greater care in their diet before calving. Cows should not be suddenly turned into rich pasture after calving, but should have some mown grass for a few days before going out. I have not had a case of parturient apoplexy since I have adopted this plan of feeding and management. Heifers and cows with first and second calves do not require this treatment, as parturient apoplexy is very rare before the third calving.

ALFRED BROWN.

The Hon. Mrs. Cecil Howard, of Dutchlands, Great Missenden, Bucks, writes as follows :—

"With regard to milk fever the treatment of the cow must depend on "the nature of the soil and on experience gained by keeping cows on the "land, since the treatment that suits cows on one sort of grass land will not "suit them on another. On the clays of the Leicestershire feeding land (where "they say bullocks can be fatted without cake) I never had a case of milk "fever, while on the gravel of Wimbledon Common I have had cases, as well "as on the chalk of the Chiltern Hills; therefore on such soils as the last two "I advise the cows to be kept off the land for six weeks previous to calving, to "be exercised daily, their temperature to be taken, and to be fed accordingly; "doses of sulphur and glauber salts should be given, varying in strength "according to the constitution and condition of the cow. No hard-and-fast "rules can be laid down. Food and medicine must depend upon circum-"stances. I do not believe in feeding on straw, but on any easily digested "light food; hay is good if not off land that usually gives milk fever. Cows "should be tethered out on bare keep if the weather is not too hot, as well as "exercised. No medicine at time of calving should be necessary or given if "the state of the cow's blood is then good. If a cow does go down, 'drops,' "then chloral and bromide of potassium given in doses as recommended by "science should be used. I have got cows up with these remedies, but I con-"sider, as a general rule, with ordinary precautions there should be no fear "of milk fever. The usual dose of glauber salts is 9 ozs., sulphur 6 ozs., in 2 "or 3 lbs. of treacle and then mixed in a quart of warm water and 1 oz. nitre "added—this last is optional—to be given a few days before calving, and twice "previously during the six weeks off grass. No bran mashes either before or "after calving for three days; oatmeal gruel after calving is better than bran "mashes, and strained linseed gruel at all and any time either before or after, "no linseed oil to be given either before or after calving."

The system adopted by Mr. G. P. Mead at The Woodlands, Bicton, Shrewsbury, is as follows :—The animals are kept on a bare pasture, with the addition of a little hay night and morning. Two weeks before calving, bran mashes, with flour of sulphur if necessary, are given as a laxative.

Eighteen hours before calving the animal is given nothing except a pint of linseed oil. In his opinion it is very important to let a cow calve on an empty stomach. After calving she must be kept warm and free from draughts. He writes:—"As soon as the calf is born, put a little salt "on it, or oatmeal, and let the cow lick it, say, for twenty minutes; then let "her drink the water off a bran mash, but on no account let her have the "bran. The mash should be made with boiling water, well stirred up as the "water is being poured on, then allow the bran to settle for a few minutes, "pour the water off into another bucket, cool it down and give it to the cow. "She may have one of these drinks three times a day. If the cow should "show signs of milk fever she should have a pint of whisky at once, follow-"ing it, if she were down in four hours, with another pint, and in another "four hours with half a pint. Rub her back with camphor liniment, keep "the rugs—which should be of wool, not jute—over her, and work with your "arms underneath them. It is very important to keep her warm and in a "natural position, well up with straw, and change her side every three hours, "or she will have paralysis; also rub her legs. She must not be left a "moment, day or night. I may add I have never yet lost a cow with milk "fever, although they have been in a collapsed condition for seven days, and "I have calved them successfully afterwards. Aconite is useful but affects "their milking. For the first seven days after calving, the cows get nothing, "but the bran water and green hay, 6 lbs. in twenty-four hours. If they do "not go down with milk fever in forty-eight hours they should not with "skilful management go down at all, but I myself exercise the greatest "caution for ten days. It should be borne in mind that if a cow goes down "with fever and gets up again, she may go down again if improperly fed "during any time up to six weeks." Mr. Mead does not consider milk fever contagious or infectious, but recommends that all the litter should be taken out and plenty of live lime put on the bottom of the cowshed.

Mr. W. P. Arkwright, of Sutton Scarsdale, Chesterfield, Derbyshire, takes the cow off the rich grass in summertime, three weeks before calving if it is her third calf or a later one, and keeps her in a yard. With regard to the treatment previous to calving he writes as follows:—"The use of the "clinical thermometer is all-important. $101\frac{2}{5}°$ F. is the normal temperature "of a cow, and her temperature for a fortnight before calving should be "taken daily. If it goes down, more generous food, preferably oil cake, "should be given until it again becomes normal. Of course the cow's "bowels must be kept open with purgative doses as required. After calving, "the passage of the vagina should be thoroughly cleansed with warm water "and Condy's fluid—$1\frac{1}{2}$ tablespoonfuls to a quart—and this should be "repeated about an hour after the cow has cleansed. I consider milk fever "to be a dangerous misnomer for the disease, as in my experience the tem-"perature has generally fallen during its attack, therefore I think that "previous starving and aconite and all lowering drugs are absolutely wrong. "If a cow shows signs of milk fever stimulate with whisky. After four "days I find that danger is over, but I may add that since the adoption of "my present system I have not had one case of milk fever, though previously "I lost nearly all my best cows."

In Miss Standish's herd the treatment adopted is as follows:—The cows are kept away from rich grass and other succulent or rich food five or six weeks, but get exercise in a yard or bare pasture according to the season. A week before a cow is due to calve, 1 lb. of Epsom salts is given, preferably in treacle, followed by a second dose if she goes beyond her time or is looking too well. There has only been one case in her herd during the last seven years, and then the above treatment had not been carried out. The cow was treated with aconite and belladonna according to instructions, mustard and blisters were applied to the loins and ice bags were kept on the head; a wineglassful of whiskey was given at intervals of two hours when the cow was in a state of coma. She was down for a fortnight and then paralysed for a considerable time, but recovered and has calved safely since.

In Mr. Joseph Brutton's herd, Yeovil, Somersetshire, the management is as follows:—Every dry cow has 2 ozs. of salt a day in her drinking water, and only a little hay. She must on no account be allowed to eat her bedding, for there is nothing more likely to cause milk fever than a cow calving with an extended paunch; it is better therefore to bed with moss litter or sawdust. If the cow is in high condition give two of Golledge's red drenches with 1 lb. of treacle in 2 quarts of old beer at intervals of a week or a fortnight before calving; but with judicious feeding this drench is unnecessary. "I seldom drench before calving. A few days before I expect the cow "to calve I have her placed in a loose box bedded with straw which has "been used for other cattle, or better still from stables, straw being better "for the calf than moss litter. Immediately she has calved, whether in "summer or in winter, I rug her comfortably and allow her to be so at least "four days. In every case I give either one of Day, Son and Hewitt's or "Golledge's red drenches in 2 lbs. of treacle and 2 quarts of old beer; after "four hours, if I have any suspicion that the cow is likely to have milk "fever, I commence the latter veterinary's treatment, which I can thoroughly "recommend. I have had several cases under my notice of cows being down "for two days and getting quite right again. I have also tried Moore's "homoeopathic treatment with much success. The calf should remain with "the cow for at least ten days, and on no account whatever should the cow "be allowed to be milked until after the second day."

In the Earl of Warwick's herd at Easton Lodge, the cows six weeks before calving are kept on nothing but good chaff or hay, and given, if necessary, a laxative drink to avoid constipation. The following treatment, recommended by the late Mr. George Simpson, of Wray Park, Surrey, is always adopted:—"Directly after calving give a purgative drink. Two hours "afterwards give 10 drops of aconite in 2 ozs. of water, repeating the dose "in an hour; two hours after that 10 drops of belladonna in 2 ozs. of water "and continue to alternate the aconite and belladonna every two hours for "twenty-four hours, after which every four hours, and on the third day "discontinue the medicine." Immediately after calving, the cows have a warm bran mash, three or four hours after the same again, then a little bran and hay, and continue this food for four days; a few crushed oats are added after this until the sixth day, when the usual food is returned to.

Mr. W. Adams, of Gloucester, for five weeks previous to calving gives nothing but hay and good straw. He considers grass and bran mashes are a pernicious evil at that time. The cows are exercised daily. Six weeks before calving he gives 6 ozs. of sulphur and 1 oz. of nitre in a quart of warm water with treacle added, and repeats this every fortnight, the last dose being a day or two before calving. Immediately after calving the following drink is given: 8 ozs. of glauber salts with a little treacle in a quart of water. Mr. Adams relies entirely upon the above treatment with dieting and regular exercise before calving, and although he has always had deep milking dairy cows on rich land he has never had a single case of fever.

Mr. Herbert Padwick's treatment a month before calving is as follows:— Keep the cattle on straw or very rough hay with just enough mangel to prevent constipation, with a small dose (½ lb.) of Epsom salts with a little ginger in it, about a fortnight and then again about a week before the cow is due to calve. If constipated, a little treacle occasionally. Since adopting this treatment, that is for the last thirteen years, he has never lost a cow from milk fever, though before that time he was a great sufferer from this preventable scourge. The cows are a little weak at calving when thus treated, but rapidly gain strength on a liberal diet and yield their maximum quantity of milk. If a cow shows signs of milk fever, his treatment is as follows:—Administer croton oil and Epsom salts immediately, and as soon as swallowed bleed hard until the pulse falters; get the bowels open in some way, using clysters frequently if necessary, and rub mustard well into the back along the spine and over the loins. Mr. Padwick considers milk fever contagious, but not infectious at the time or shortly after calving.

In the Dowager Lady Freake's herd at Fulwell Park, Twickenham, the cows are kept in the straw-yard during the two weeks before calving, and are fed on hay only and no succulent food whatever. Two days before they are due to calve, the following drink is given:—1 lb. of glauber salts, 1 teaspoonful of ground ginger dissolved in 1 quart of hot water. This dose is repeated if the cow goes many days over her time, and also after calving with the addition of 1 lb. treacle. In cases of heavy milkers 10 drops of aconite in a wine-glass of water three times a day, as a preventative, for a few days after calving. Only two cows have been lost from milk fever in twenty-five years, when the above treatment was not strictly followed.

Mrs. Peel, Byletts, Pembridge, Herefordshire, who has kept Jerseys since 1875, advises that the cows should be kept very short of food for six weeks previous to calving and not allowed out in the fields but should have plenty of walking exercise, and drinks should be given to them three weeks before they are expected to calve. The drink used in her herd is as follows:—½ to ¾ lb. of Epsom salts, 1lb. coarse brown sugar, 2 ozs. of ground ginger, in warm water, one dose two weeks before calving and a second dose a week before calving.

In Mr. C. Combe's herd at Cobham Park, Cobham, Surrey, if a cow shows signs of milk fever she is treated as follows:—1 pint of whisky is given every three hours, a strong liniment is rubbed in over the loins, a fresh sheep skin is put over the back, care being taken to put the flesh side inwards, and ice is applied between the horns.

Mrs. Burton, of Longner Hall, Shrewsbury, who has kept Jerseys for over twenty years, puts the cows on a closely grazed pasture and starves them at night for six weeks previous to calving, and for a fortnight before calving gives a quart of linseed tea every day; three or four days before calving the following drink is given: Nitrate of potash 1 oz., sublimed sulphur 4 ozs., pulv. nux vomica 30 gr., valeria 1 oz., dia pante 1 oz., Epsom salts 12 ozs., mixed in three pints of warm water. This is repeated as soon as the cow has calved, and if the case appears suspicious it is repeated twelve hours afterwards. The cow is fed on hay and bran mash. This treatment has saved the lives of many cows.

Dr. Alcock, Byne House, Warminster, Wilts., recommends that previous to calving the animal should not be over-fed, and only be given food which keeps the bowels fairly open. He does not give any medicine beforehand. If a cow shows signs of milk fever he blisters the spine from the horns to the roots of the tail with mustard, and gives half a pint of whisky every four hours. He has never lost an animal since he adopted this treatment.

In Mrs. Thackwell's herd the treatment in a case of milk fever where the cow is very fat is to bleed her, taking 2 or 3 quarts of blood and giving plenty of whisky. For aperients, 1 pint of linseed oil and 15 to 20 drops of croton oil is recommended, as well as injections of warm water and soap. The following drink should be given for two hours regularly: 2 drachms of bromide of potash, 2 drachms of chloral hydrate, ½ lb. treacle. After this dose purgative medicine need not be given.

In the Duke of Grafton's herd, at Wakefield Lodge, Stony Stratford, Bucks., the cattle, if in high condition and heavy milkers, are kept on short diet and with plenty of exercise, mild doses of Epsom salts and treacle, or 1 to 2 ozs. of hyposulphite of soda are given occasionally. In winter the cows have a small quantity of linseed cake, as no roots are used. This treatment commences a month before calving. After calving, if the cow shows decided signs of milk fever, the following is given:—Epsom salts, 16 ozs.; bitter ergot of rye, 1 drachm; nux vomica, 1 drachm; sweet nitre, 8 ozs.

Mr. Thos. Le. Sueur, of Hyde End Farm, Brimpton, Reading, keeps his cows on dry food, hay and straw six weeks before calving, with as much water as they like to drink, and plenty of exercise. He gives two purging drinks before they calve (eight days interval) and one as soon as they have calved. The dose he recommends is one pint of castor oil. If the cow shows signs of milk fever he gives a pint of Irish whisky, and one pint of linseed oil, the cow being rugged up. Two hours afterwards 15 drops of aconite and 15 drops of belladonna are alternatively given. Mr. Le. Sueur would whitewash and disinfect any cow-house where a cow has been down with milk fever.

CHAPTER V.

ABORTION; PREMATURE CALVING; STERILITY.

THERE is perhaps no disease so dreaded by the breeder of stock as abortion. The suddenness with which it comes into a herd, and the pecuniary loss entailed are bad enough, but the uncertainty as to where it will end, or how it is to be stamped out, causes the greatest anxiety to the owner of a valuable lot of cattle.

The subject of abortion has been discussed by almost every writer on cattle pathology, and by none more clearly than Professor Sir G. T. Brown and Mr. Clement Stephenson in their articles on "Abortion in Cattle," in the *Journals of the Royal Agricultural Society of England* for 1885 and 1891, where the two forms of abortion, "sporadic" and "epizootic," are defined and the treatments recommended set out concisely and simply. Professor Bang, of Copenhagen, has since established by experiments what the previous writers had considered probable, that epizootic abortion ought to be regarded "as a specific uterine catarrh" determined by a definite "species of bacterium." This "uterine catarrh which is set up by "the abortion bacilli does not always entail the expulsion of the "foetus, the result being sometimes the death of the foetus only." The most important discovery, however, is that the "Bacilli "retain their vitality for at least seven months, and are not expelled "of necessity with the foetus, but remain in the uterus unless a "careful disinfection of the uterine cavity is carried out. The "vitality of the bacillus also explains the difficulty encountered in "getting rid of the disease from an infected building."[1]

With a view of ascertaining the opinions of, and the treatment adopted by, the breeders of Jersey cattle who had suffered from this scourge, the various questions set forth in the preface were asked, as it was felt that even though little might be added to the knowledge of the disease, the remedies tried might at any rate confirm the suggestions which have been given in the three articles above referred to.

---

[1] *Bath and West of England Journal*, 1897-8, p. 220.

Abortion has been defined as "the premature expulsion of the "impregnated ovum, the embryo, or the fœtus before vitality," and premature calving as "expulsion of the fœtus after vitality." Sterility frequently follows both forms of the disease. Sporadic abortion, which is by far the commoner form, is caused by the following :—(1) Bad food of various sorts, and impure water; (2) hot, dirty, and badly constructed stables with inside drains; (3) blows, strains, slips, falls; (4) fright, excitement.

In properly managed herds this form of abortion should be rare, as it should not be traced to preventable causes, such as musty food, impure water, badly constructed or dirty buildings. In like manner, blows, strains, slips, falls, and other mishaps should not occur if the cowmen take ordinary care in bringing the animals in and out of the cow shed and meadows, and also see that cows in use are not turned out with the others.

Sudden frights, which are very often the cause of this form of abortion, are of course sometimes unavoidable; and to prevent or mitigate these evils as much as possible, is one of those traits of cleverness and good management which distinguish a good from a bad cowman.

Abortion in heifers is said to arise from want of sufficiently nourishing food, and as the danger of milk fever with them is almost *nil*, there is no need to starve them during any part of their pregnancy.

As in milk fever, prevention is better than cure. When any premonitory signs of abortion are noticeable, the cow should be removed at once to a separate building so that the disease may be kept within bounds.

The fœtus and excreta should be burnt or buried in quicklime. If the abortion has taken place in the field, the ground should be dressed with quicklime and hurdled off, the hurdles not being removed until the top dressing of lime has disappeared. If the abortion takes place in the cow sheds, the standing and immediate surroundings of the animal should be thoroughly disinfected.

Epizootic abortion is infectious, and comes from contact, or sympathy induced by contact, with animals that have aborted, or with the discharge from these animals. The treatment adopted amongst breeders in cases of this form of abortion varies. In some herds Mons. Nocard's system is adopted, in others antiseptic injections are recommended. In a few cases internal doses of carbolic acid have been found most beneficial; while in some obstinate attacks the old fashioned remedy of a Billy-goat has apparently succeeded where all other remedies appear to have failed.

The system adopted by Mons. Nocard described by Professor Sir G. T. Brown, in the article mentioned above, is as follows:—

[1] (1) The weekly disinfection of the cowstalls and drains behind the cows with a solution of sulphate of copper, phenic acid,—*i.e.*, carbolic acid, and corrosive sublimate.

(2) A daily washing, by means of a sponge saturated with the solution of corrosive sublimate, of the anus, the vulva, the perinæum and the tail of all the pregnant cows. The prescription used for this purpose is as follows:—

Rain water about	10 pints.
Corrosive sublimate	2½ drachms.
Hydrochloric acid	1½ ounces.

It is recommended that this treatment be kept up for a year at the least, but in the light of recent discoveries by Professor Bang, it does not seem to go far enough. Antiseptic injections or doses of carbolic acid, which latter, according to the experiences of one or two breeders, seem to have a great effect on the uterus—should form part of the treatment, in fact, they should be a complement to that recommended by Mons. Nocard.

The doses of carbolic acid are as follows:—

Two teaspoonfuls of Calvert's No. 5 acid are mixed in a pint of water, and poured into a scalding bran mash, which must have been made of the best English bran, as otherwise it will not get into the necessary glutinous state. The bran mash must be well stirred until nearly cold, when it can be given to the animal. This dose should be given every day for a week.

The general opinion appears to be that premature calving and abortion only temporarily affect the milking properties of the cow, and that if she subsequently calves at her full time, the milk yield will be as good as before. Where a cow has calved prematurely and still remains in the herd, she should not be sent to the bull

---

[1] See "Abortion in Cattle," *Royal Agricultural Society of England Journal*, 3rd series, vol. ii., part 4, 1891.

until two or three months after the time she should have carried her calf, and then only if she has thoroughly cleansed, as otherwise she may infect the bull.

The experience of those who have had cases of abortion in the herd points to the conclusion that a bull, after service of a cow that has calved prematurely or aborted, may infect other cattle. Professor Bang's discovery of the bacteria appears to strengthen this opinion.

Sterility frequently, though not invariably, follows cases of premature calving or abortion. The bacillus may account for this, as the periodic returning to the bull is by some veterinary writers considered as showing that an immature abortion has taken place. Temporary sterility may arise in young animals from various causes, but over-feeding, both of the male and female, is generally considered by breeders as most conducive to sterility. Several remedies are suggested—reduction of food, bleeding, and treatment similar to that recommended for abortion, but unless the animal is a very valuable one, the better plan, and that generally adopted, is to get rid of the non-breeder at once, since the presence of such an animal is prejudicial to the rest of the herd, in more ways than one.

Notes on Abortion, Premature Calving and Sterility, received from Breeders.

The Hon. Mrs. Cecil Howard, of Dutchlands, Great Missenden, Bucks, writes:—" The dose I give for prevention of cows catching abortion is
" daily for a week ¼ oz. Calvert's No. 5 carbolic, shaken in a bottle of water
" and mixed in a *large* pail of boiling hot bran mash; it should be well stirred
" and given when cool enough. This quantity of carbolic can only be taken
" if the bran is in a glutinous state, and cannot be made with foreign or
" inferior bran; in that case a mixture of carbolic and oil should be
" obtained from the veterinary surgeon and administered as he directs.
" Each cow to be fed separately, and each mash to be mixed separately and
" fed in separate troughs or pails. This dose can be repeated at intervals
" of a week apart if several cows abort, as it will prevent the disease if the
" cows are not already infected. Naturally the cow that had aborted would
" not be with the others, nor should the cows be turned on the same pasture
" as the one where the abortion may have taken place. The carbolic acid
" does not affect the taste of the milk in any way as long as the milk pails
" and milk are not left standing in the cow sheds. The pails should be taken
" out of the shed directly the cow is milked."

Mr. J. I. Thornycroft, of Steyne, Bembridge, Isle of Wight, considers that contagious abortion is caused by a bacillus, and is a distinct disease to those cases which are caused by fright, &c. He mentions a case in the Isle of Wight which came under his knowledge, which is so interesting that it is given *in extenso*. "In one case where a herd of Guernseys were taken to "a farm where there had been an outbreak of contagious abortion, a large "number of the cows aborted, presumably from infection in the cow sheds, "as all the cattle belonging to the previous tenant had been removed "before the Guernseys were taken to the farm. The outbreak was stopped "after some months by giving all the cows large doses of carbolic acid in "the food every day. The sheds were all disinfected, but the outbreak was "not stopped until the carbolic acid treatment had been continued for "some time. This occurred a year or so ago, and I believe there have "been no cases of premature calving in the herd since."

Mr. Crawford, agent to Mr. W. O. Hammond, considers abortion infectious and to some extent contagious. He writes: "After abortion a "quantity of virus is left in the uterus beyond the reach of antiseptics, and "if the cow is allowed to go to service before thoroughly cleansed (and this "can only be accomplished by time) the trouble will appear again. The de-"composed matter expelled when epizootic abortion takes place, and the "uterine discharges for some time afterwards, contain septic germs which, "under favourable circumstances, can be again introduced into the system "through the blood (not, as is generally supposed, by the vagina) and are "capable of reproducing the disease. Cases of sporadic abortion may be "caused by accident, fright, exposure, ergot, and chills, but if precautions "are taken there is little further trouble. Epizootic abortion—the dread of "all breeders—is frequently introduced into the herd by cows purchased from "affected herds; in the case of neglect and want of care in sporadic cases— "that is to say, when the cow aborts from some accidental cause and little "notice is taken of it—70 per cent. abort again when from four to six months "are gone. This time a quantity of offensive matter is expelled, heavily "charged with infectious germs." The treatment Mr. Crawford recommends is to thoroughly inject a mixed antiseptic into the vagina and in every case to isolate the cow.

Mr. Richardson Carr has tried several remedies, but cannot say with certainty whether they have been successful, because in his experience abortion goes as quickly as it comes. A pint of crushed hempseed, injections with disinfectants, doses of carbolic acid, and a billy-goat, have each, in their turn, appeared efficacious. He recommends the following medicine and wash as a possible prevention to abortion in cows:—

Internal Medicine. Carbolic acid pure, 1 drachm; glycerine, 2 drachms; water (warm), 3 ozs. Dissolve the carbolic acid in the glycerine and then add the water. The above makes three or four doses, according to the age and size of the cow, to be given once a day, but in severe outbreaks of abortion it may be given twice a day, for a time at any rate, and in the doses mentioned it does not affect the milk.

The wash for *external* use is as follows:—Bichloride of mercury, 2½ drachms; hydrochloric acid, 1½ ozs.; rain or distilled water, 10 pints. Apply daily with sponge to vulva, anus, and under side of tail.

Mr. J. F. Hall writes:—"I believe premature calving to be one of the evil "effects following frequently upon abortion. As to the causes of abortion "they are numerous and obscure: sometimes it seems to be an epidemic; "sometimes to be introduced by serving neighbours' cows with the bull used "in your own herd; sometimes it is mechanical, as when caused by a blow "from another animal; sometimes it can be traced to indigestion, such as is "occasioned by overfeeding in winter with cold watery roots; sometimes it "accompanies a consumptive habit of body; sometimes it is ascribed to "ergot in the grass or fodder, and sometimes—I am inclined to think more "frequently than is generally supposed—it is due to lack of nourishment at "some period of the cow's pregnancy. It prevails much more in some "districts and in some seasons than in others. During eleven years of "Jersey farming in Berkshire, I only met with two or at most three cases in "my herd, whilst during eleven years in Somersetshire I have seen a great "deal of it both in my own herd and elsewhere. In 1887 and 1888 I killed "six of my best cows in the hope of stamping out the disease, but was only "partially successful. The only treatment I have found efficacious is one "of slow action and is described in the *Royal Agricultural Society of* "*England's Journal, 3rd series, vol. ii., 1891*, it is called Nocard's method. "If persevered in I am of opinion that this treatment is effectual."

In answer to question 59, "How long after abortion or premature calving do you allow the cow to be served?" Mr. Hall says it depends on the health of the cow, but as a general rule about six weeks after the date at which she would have calved had she gone her full period.

Mr. Hall's experience of temporary sterility has been very exceptional. In one year out of 120 services 86 or about $72\frac{1}{2}$ per cent. proved unavailing. It followed upon a very dry spring and summer when grass was very short and the cattle got low in condition and possibly lacked some of the sustenance required during some portion of the pregnant period. This may have occasioned weakness leading to the expulsion of the fœtus. If it was not this it must have been in the nature of an epidemic. He recommends a change of bull and blood-letting in cases of sterility, but adopts Nocard's solution and treatment as in cases of abortion.

The treatment in Sir James Blyth's herd is as follows:—The after-birth is removed immediately it falls, and is buried, so as not to cause excitement or sympathy amongst neighbouring animals. Directly the least symptom is detected, such as an unnatural discharge, any increase of size behind, or swelling of the udder before the animal's natural time for calving approaches, she is removed from the herd and isolated, and no contact is allowed with the rest of the herd or with the man attending her. Abortion is thought to damage the milking properties of cows for a time, although not permanently. With regard to sterility, the experience is that it varies with the different natures of the animals; some will hold to service if fed and got into high condition, while with others the very opposite treatment will be found successful. Throwing cold water on the back immediately after service has sometimes been attended with good results.

The treatment in Earl Cadogan's herd at Culford is as follows:—As soon as the cow shows any signs of abortion, she is removed from the others and given laxative medicines, just sufficient to move the bowels. Corn is stopped and hay only is given. Low's medicines (Fred Low, M.R.C.V.S., Norwich), have been tried twice and in both cases successfully.

Mr. Padwick attributes cases of abortion in his herd—Firstly, to the cows drinking water that has been rendered brackish by overflow from the sea or leaking sluices. He considers this a certain cause of abortion if continued many days together, as the cattle will drink this water if they can get access to it, even though abundance of fresh water be provided; Secondly, to animals being kept too low in condition, especially in the case of heifers. Thirdly, to excitement from being driven about by dogs; jumping ditches and running up and down the sea banks. Isolation and the use of disinfectants have been tried and found to give perfectly satisfactory results.

Mr. Padwick attributes *sterility to very high feeding*; his treatment is to *reduce the artificial food* and let the bull run *with the cow*.

Colonel Willan writes: "A few years ago, when about twelve or fourteen "cows were lying one night in a yard to harden them in the spring, before "lying out in the field, one of them aborted. They started aborting at all "ages up to seven or eight months' gone, and this continued for eighteen "months, or nearly so. I tried every nostrum I could hear of and every "device as to service, disinfectants, &c., the result being 'blank despair.' "As a last resource an old Billy goat was bought for twelve shillings, and "abortion was stopped at once. A goat has run with the cattle ever since, "and now they are practically never troubled with abortion." Col. Willan considers the goat to be nothing but a strong ever-present natural disinfectant.

Mr. T. Loader Brown mentions an exceptional case of long-continued sterility in one of his best cows. "She was ten years old and produced a calf "on January 27, 1895; she was served the following May 15, but returned "June 8, and continued returning until July 6 of the following year, "having been served eleven times by various bulls, and allowed to pass at "other times; she held at the eleventh service and duly calved in April of "the following year, 1897. Being a prize butter test cow, with a record of "2 lbs. 9¼ ozs. per day, she was kept as above and gave a large quantity of "milk for more than two years after January, 1895. She has subsequently "bred well. During the whole of the above period she appeared in perfect "health."

Many other breeders take the same view of the value of the goat as a disinfectant, and allow goats to run with the herd. One breeder in Gloucestershire[1] states he had had about twenty cases of abortion up to the year 1888. He then let an old he-goat or Billy goat run with the in-calf cows and heifers, both before and after they went to the bull, and he has never had another case of abortion and the heifers have bred regularly since.—ED.

---

[1] Mr. W. E. Budgett, Henbury, Bristol.

The following forms of Treatment and Recipes for Abortion and temporary Sterility have been received.

Mr. T. Clarke, Steward to the Dowager Lady Freake, Fulwell Park, Twickenham.—In cases of temporary sterility the treatment adopted is to give one pint of crushed hempseed once a week in a bran mash; this has been attended with good results. The cow is not sent to the bull until she is going out of heat.

Mr. L. G. Gisbourne, Allestree Hall, Derby.—Isolation of all affected cattle for three months, weekly cleansing of all sheds by whitewashing and sprinkling with carbolic acid solution, daily sponging the vulva of healthy cattle with a solution of carbolic acid in water, and administration of ½-oz. carbolic acid in bran mashes twice a week. This treatment has met with partial success as it is almost impossible to get complete isolation long enough. Outbreaks generally continue at intervals for three years, when the disease seems to gradually wear itself out.

Mr. E. Murray Ind, Coombe Lodge, Great Warley, Essex.—In a case of abortion the cow should be separated from the rest of the herd and the other cows in the herd should receive small doses of carbolic acid in bran mashes every morning for a week.

Mr. Thos. H. Lukes, St. Austells, Cornwall.—Isolate every case and freely use disinfectants; remove all manure without delay, using cleansing drinks. After this give the cows carbolic in the powder ½-oz. the first week increased to 1 oz. up to four weeks; then reduce again for another week. This treatment seems to purify the womb, but the cow must not be put to the bull for a month after the treatment is stopped, and this medicine must not be given to her after going to the bull, as the treatment with carbolic powder may cause abortion, since it acts very strongly on the uterus.

Mr. W. Shipton, Ivy House, Scropton, Derby.—In cases of sterility two quarts of blood should be taken from the animal at the time of service and a dose of ¼-lb. of Epsom salts and 1 lb. of treacle be given two hours before service.

Mrs. Thackwell, Rostellan Castle, County Cork.—Doses of carbolic acid, commencing with ½-drachm and increasing to ½-oz. given in new milk until the cow calves.

CHAPTER VI.

Calf-rearing.

The feeding of Jersey calves, like the feeding of Jersey cows and heifers, is peculiar to the breed. They require special treatment for the reason that, being wanted for the dairy, they must not be allowed to get fat, but be kept in a healthy growing condition.

In Jersey, cake is not given to heifer calves, and although in this country the climate is not so good as it is on the Island, yet the Jersey system of calf rearing should be followed to this extent—that no fat-forming foods should be given to those animals that are intended to be kept for the herd.

In the chapter on sterility the opinion of many breeders is that those animals which are difficult to breed from have been over-fed; this alone should demonstrate that cattle for the dairy must be fed sparingly on cake and fat-producing foods, with a view of developing the milk-producing and not the fat-producing tendencies in the animal.

The system of calf rearing generally adopted is as follows :—

The calf in most cases is allowed to remain with its dam, if from a heifer, for ten to fourteen days, if from a cow, for two or three days. After being taken from its dam it is fed upon new milk for the first three weeks or a month, then separated or skimmed milk is added, until the new milk is gradually displaced. In some cases, calf meal, in others, boiled linseed is mixed with the separated milk, the quantity being gradually reduced until the diet is reduced to separated milk only. This should be discontinued when the calf is six months old.

As soon as the calf can nibble, a little hay is given with small quantities of crushed oats and bran. Rock salt and a lump of chalk in the manger are recommended. When old enough the calf may be fed in winter on pulped roots and chaff.

In the summer, calves six months old and upwards may be turned out to grass in a meadow away from the cows, and rock salt should always be within their reach.

The same remarks as to feeding heifer calves apply to bull calves; over-feeding is a cause of sterility, and the best foods for bulls are farinaceous, such as oats and bran.

In rearing calves the following are of paramount importance—cleanliness, regularity, careful attendance.

The calf shed should be kept clean and sweet, the pails from which they drink, whether ordinary or Tucker's, should be washed with boiling water each time they are used (this especially refers to the indiarubber teats in the Tucker's feeding pails), and the mangers or boxes which contain the other foods should be cleaned out daily.

The feeding should be at regular intervals, the milk given should be at an even temperature, and the amount of food, both milk and other sorts, should be properly measured.

A careful calf rearer will always see that the animals are kept warm and brushed over when young, remembering that calves miss the attention given them by their dams. For this reason, they should always have their mouths and nostrils dried after drinking their meal or milk.

If these few points are attended to, calves should do well, but the neglect of them is apt to cause scour, indigestion, and other ailments which should not occur in a properly managed herd.

In reply to the question as to the treatment of scour various remedies are given, and the answers which are given below and in the following pages should be consulted. To prevent scour, the addition of lime water to the milk is strongly recommended. When an attack of scour commences, the bowels should first be relieved by a small dose, castor oil being the most effective, followed by such remedies as are suggested on the next page, and the calf should be kept warm. What these are will best be seen from the extracted replies, those in most general use being doses of linseed or castor oil, bicarbonate of potash, new laid eggs, laudanum and chlorodyne.

Notes on the Rearing of Calves received from Breeders.

Colonel C. P. Le Cornu, of La Hague Manor, St. Peters, Jersey, who has been President of the Royal Jersey Agricultural Society on several occasions, and a breeder of Jerseys for over forty years, states that his practice is to take the calf away from the cow directly after calving. If the calf is not to be fatted for the butcher it is given sweet milk for the first three weeks, after which it is gradually fed on skim milk and a little hay. In the summer and autumn, when the calves are old enough, they are turned out in a paddock with a shed for shelter, and in addition to the food mentioned they are given grass. No farinaceous or artificial food is given.

The Ladies E. and D. Hope, Big Hollanden Farm, Sevenoaks, recommend that the calf be always taken from the cow directly she has calved, as the cows milk better and the calves are more easily reared if the separation takes place at once, there being no fretting on either side. It is, however, essential that the calf gets its own mother's milk for a few days *at least*. Since adopting this plan they have never lost a calf. If the calves show signs of scour, the milk allowance is decreased to half the daily quantity, and a new-laid egg is beaten up in it.

In the Blythwood herd the calf is left with the cow for three or four days, so that it may get the colostral milk nature provides. The calves are tied up so as not to suck one another, and are fed on warm separated milk in which a little well-boiled linseed has been put to replace the cream. They are fed afterwards on dry bran and crushed oats.

In Mr. Hall's herds, at Sharcombe and Chilcote, the calves are left with the dams for three days, and in the case of a heifer rather longer. Mr. Hall considers the best system of rearing calves to be as follows: first, leaving them with their dams; next, weaning them on scalded skim milk— that is, milk separated from cream by scalding; third, by use of hay tea or of milk separated from cream by mechanical means. The larger number of calves in his herd are weaned and reared on separated skim milk to which a certain proportion of new whole milk has been added. This is warmed whenever required. Each calf receives from one to two gallons according to age and size. They get also a little long hay and linseed meal directly they are old enough to eat it. The milk is given twice a day.

Mr. Padwick writes:—"My calves remain with the cows for a week; "they then have new milk for a fortnight from the time of weaning; this "is gradually mixed with separated milk for another fortnight, then "separated milk with a little boiled linseed added. As soon as they eat "well enough to do without milk they run loose in barns, and have up to "six months old the same mixed food that the cows are having, and as "much of it as they will eat up clean."

In the herd at Tring Park the treatment is as follows:—If the heifer is very excitable, the calf is left a few days. The calves receive a quart of new milk night and morning for the first two months, they then get a little gruel, with half the quantity of milk, up to six or seven months. In the event of scour, Sutton's (Norwich) calf drinks are used.

Remedies for Scour.

The following forms of treatment and remedies for scour have been received from members.

Doses of opium and ammonia have hardly ever been known to fail.—Mr. Arthur Arnold.

A teaspoonful of Gregory's powder in a cup of warm water; six hours afterwards a little magnesia and peppermint.—Blythwood Herd.

Castor oil and a mixture of prepared chalk and bole armeniac.—Admiral the Hon. T. S. Brand.

Oil is given in the first instance and sometimes bicarbonate of potash to correct acidity.—Col. Le Cornu.

One or two tablespoonfuls of cold drawn linseed oil; an hour afterwards a teaspoonful of bicarbonate of potash dissolved in warm water.—Mrs. Custance.

A new laid egg put down the calf's throat, shell included, and no milk given for a few meals.—Fulwell Park Herd.

Dose of castor oil followed with two or three raw eggs daily. Feeding on flour gruel is sometimes efficacious, and starch balls is another remedy.—Col. Walter Hankey, St. Leonards-on-Sea.

One teaspoonful of bicarbonate of soda in the milk.—Mr. T. Jefferson.

A dose of castor oil and ginger should be given at once followed by pills of butter and bicarbonate of soda mixed ($\frac{1}{4}$ lb. butter with a teaspoonful of soda, well mixed). This has been found very efficacious, experience showing that if taken in time it rarely fails.—Mr. W. Milward-Jones.

A small dose of linseed oil followed by two to four tablespoonfuls of the following:—Prepared chalk 2 ozs., powdered catechu 1 oz., ginger $\frac{1}{2}$ oz., opium 1 drachm, peppermint water 1 pint. Fresh eggs, shell included. Lime water and old beans may often be used. A piece of chalk is put where the calves can lick it.—Earl Cadogan's Herd.

Port wine and blackberry shoots made into tea, half-a-pint each time three times a day, and old acorn powder.—Col. Barton Scobell.

Ordinary cholera mixture, giving twice the human dose.—Col. T. B. Shaw-Hellier.

Gaseous fluid, but if not obtainable, pills made of the best yellow clay and a small quantity of spring water.—Mr. W. G. Skinner.

Beat up an egg, shell and all, and mix with a little salt, or give a dose of carminative chalk in milk.—Col. Kenyon Slaney.

On the first symptoms a tablespoonful of linseed oil, followed in an hour by a teaspoonful of bicarbonate of potash in water. Repeat if necessary. If the calf is very bad and weak, an egg beaten up with a wine-glassful of port wine soon revives it.—Miss Standish.

Lime water; limit the quantity of milk.—Mr. W. G. M. Townley.

Two or three eggs with a little whisky and flour gruel, half a pint twice a day.—Earl of Warwick's herd.

One tablespoonful of castor oil, one teaspoonful of laudanum. Keep the calf warm and dry, with plenty of fresh air.—Mrs. Charles Wyndham.

Two tablespoonfuls of flour, one tablespoonful of powdered ginger, mixed into a paste with whisky, made up into small balls, and given every two or three hours, if necessary, has effected many cures in the earlier stages of this complaint.—Mr. Ernest Mathews.

CHAPTER VII.

Cost of Keep; Dairy Properties; Testing Cattle; Barreners.

A QUESTION very often asked breeders of Jersey cattle by those who have had no experience of the breed is "Do Jerseys pay?"

It is obvious that to give a correct answer the following points must be considered:—

1 Average initial cost of the animals.
2 ,, cost of keep.
3 ,, cost of labour.
4 ,, weight of butter or milk sold and the price.
5 ,, weight of separated milk sold and the price.
6 ,, value of the calves.
7 ,, value of the barreners.

No. 1, 6 and 7 may be dismissed at once from the calculations, as they must necessarily vary in different herds; the only items therefore left to be dealt with are the cost of keep and labour on the one side, and the sale of dairy produce on the other. From the answers to the questions asked on these subjects, it will be seen that a satisfactory balance sheet ought to be made out by those who keep Jersey cattle.

The general opinion appears to be, that the butter qualities in particular families of Jerseys are transmitted by inheritance, and therefore to breed cattle without considering these qualities must be wrong. In purchasing Jerseys, the pedigree, character of the ancestry on both sides, and the milk and butter record of the dams and grandams should be looked into, as it is only by taking this trouble that the breeder can expect to mate his animals properly, and so obtain good dairy strains.

Testing cattle for butter is still in its infancy; milk records, though kept occasionally, are not nearly so general as they should be. As a rule, the trouble involved is the excuse given for their omission, and perhaps the excessive zeal of a few may be the cause of the neglect of the many.

The labour of keeping milk records and testing for butter is generally exaggerated. For all practical purposes, where butter only is made, it will be sufficient if the morning's and evening's milk is weighed one day in the week, though, of course, if the milk is weighed daily it is preferable. The weight of milk should be entered into a book, each page of which should be divided into columns to take a calendar month. The total of each day multiplied by 7 will give an approximate average weight of milk per week in pounds; these can be reduced to gallons if necessary by dividing by 10, as, for ordinary purposes, 10 lbs. of milk may be taken as equal to an imperial gallon.

The amount of butter made each week should also be entered in the book, and after deducting any new milk that has been sold or used in the house, the number of pounds of milk, divided by the number of pounds of butter, will give the butter ratio of the herd. This, in a good herd, should be from 17·00 lbs. to 17·75 lbs., provided that cream separators are used, that care is taken to prevent any waste, and that the churning is carried out on scientific principles.

The butter ratio figures are arrived at by dividing the number of lbs. of milk by the number of lbs. of butter.

Dividing the number of pounds of butter by the number of cattle in the herd (and for this purpose all cows that have had a calf, whether in milk or dry, should be reckoned), the average weekly yield of butter per head per cow is obtained.

In a good herd this should never be lower than 6 lbs. per head per week throughout the year, and with careful breeding this figure may be exceeded. Hitherto only the average returns of a herd as a whole have been considered, but it is possible, that even where there is a high average, there may be one or two cows that should be drafted. Individual tests of cattle are therefore necessary, and the more so, as very often the heaviest milkers at first are not the most profitable where butter only is sold.

Each cow should be tested five to six weeks after calving, and again three months later. The object of the two tests is to gauge the lactation period, and so see what each cow is worth. In taking these tests, the milk of twenty-four hours should be weighed, separated, churned, and made up into butter, the ratio worked out, and a record kept of the two trials.

If chemical tests are preferred, the Gerber Butyrometer and the Babcock Tester are very valuable, and can be worked to a nicety, but it is essential that the acid and alcohol should be of the right strength, according to the printed instructions.

Another plan, which has been suggested, to show at a glance the period of lactation in cattle, is well worth consideration, viz., to keep a chart showing the weekly milk yields of each cow. The chart is divided into spaces by vertical and horizontal lines. The vertical lines give fifty-two spaces for the weeks of the year. The horizontal thirty-eight or forty, which represent the number of gallons yielded per week from one up to thirty-eight. The horizontal lines are again ruled off into divisions at the 7th, 14th, 21st, 28th and 35th lines, which marks the yield at one, two, three, four and five gallons per day. The annexed example, which has been kindly sent to the Society by Mr. Thornycroft, of Bembridge, Isle of Wight, explains itself.

The value of testing cows cannot be over-estimated, and if systematically persevered in, the dairy qualities of a herd will soon improve, and the breeder will learn from examining his good cows, many characteristics peculiar to butter-producing animals.

Where milk alone is sold, the keeping of records is simpler, as probably the milk is weighed or measured both morning and evening; but even then, it is well to enter in a book the individual produce of each cow once a week, to see whether she keeps up her yield, and also to record the period of her lactation.

In reply to the questions as to the annual average weight of butter given by Jersey cows under and over five years old, the answers give the following average yields per annum:—

Cows under five years, 260 lbs.

Cows five years and over, 320 lbs.

No questions were asked as to the price obtained for butter or milk, and therefore any estimate based on the value of these articles must be more or less a matter of conjecture. When the milk from Channel Island cows is sold, it should, according to the analyses of milks taken during several years at the London Dairy Show, be worth from 40 to 50 per cent. more than that of the other dairy breeds. The butter, in the same way, if properly made, should command the highest price in the market. It is incontrovertible that the butter from Channel Island cattle that are properly fed not only requires no colouring matter, but is firmer, of better quality and flavour than any other butter; and it is believed that, if well-made, it should sell at an average of 1s. 3d. to 1s. 7d. per lb. at the farm all the year round.

From the foregoing figures as to weight and price, it will be seen that a Jersey cow can give a good pecuniary return per annum in butter alone; if in addition to this the bye-products, *i.e.*, separated milk and butter-milk, which have always a market or feeding value, are taken into consideration, the net profit on each cow will be augmented.

With regard to the cost of food and labour, the replies sent in naturally vary. Labour differs in different parts of the country, and the management in some herds is more expensive than in others; but, taking the average, it would appear that the cost of attendance on a Jersey cow for a year is about £3.

The cost of keep is also difficult to ascertain. In the questions rent was to be excluded, but as the cost of the food was to be taken at market price, the rent must be included in the price of the foods. The replies show that the average cost of food, including the summer grazing, is about £11 10s. per annum.

With ordinary luck, a Jersey cow should pay for herself over and over again with her calves and dairy produce during her lifetime. An aged milking cow, whether a Jersey or any other breed, if she has ever been worth anything in the dairy, will only sell for a low sum to the butcher, owing to her peculiar conformation, and to the superior quality of young beef cattle now coming into the country from North and South America.

Jerseys which are sold for anything like a fair price to the butcher are evidently young animals which have turned out either failures in the dairy, or barren, and have then been fattened. The average price at which barreners are sold is about £11 15s., but in most cases breeders are of opinion, that a Jersey when past work is not worth troubling about. Steers are fattened by a few breeders, and, where there is a market for them, realise an apparently remunerative price.

Notes on the Dairy Properties of Jerseys received from Breeders.

Mr. J. F. Hall believes that butter properties are transmitted by inheritance in particular families of Jerseys. He writes: "Careful "observation, extended over many years, has convinced me that where these "properties exist in any unusual degree they are distinctive of individual "cows and are transmitted by inheritance. Of course they are not "invariably so transmitted. The dairy merits of the family ancestry on "the bull's side have likewise to be taken into consideration, and where "these are wanting the superior qualities of the dam may not be reproduced "to the full extent in the offspring." Mr. Hall has kept milk records and butter tests continuously for fifteen years. His system of taking them is as follows:—One day in each week the milk given by each cow morning and evening is weighed separately and recorded in a book kept for that purpose. The total multiplied by seven gives the yearly yield. To check this with accuracy, each day the bulk of the milk drawn is weighed and recorded. The two totals, viz., that of the one day's test multiplied by seven, and that of each day's bulk drawn should agree, and they are found to do so sufficiently near for all practical purposes. Mr. Hall takes periodical butter tests of the general quality of the herd, and also of individual cows.

In Lord Rothschild's herd at Tring Park, the milk of every cow is weighed morning and evening; the total milk from all the cows has to agree at the end of the week with the actual amount of milk sold. Two accounts of the milk are kept, one yearly from Michaelmas to Michaelmas—looking upon the cattle as milking machines; the other as a pedigree record, giving the details, with each calf. Butter tests of the general quality of the whole herd are taken every week, and tests of particular cows are taken twice during the period of lactation.

Miss Standish writes:—" Where butter is specially wanted, tests are most "valuable, as often a cow may appear to be only a fair milker, but the milk is so "rich in fat that she may be more valuable than one giving a larger quantity "of poorer milk."

Mr. J. I. Thornycroft's system of keeping milk records is as follows:— "The milk of each cow is weighed separately morning and evening; the "weights of the morning's and evening's milk are added together and entered "in a book each day and the entire weight of milk from the time the cow "comes into the shed until she dries off is added up at the end of her milking "period. Recently we have made *curves* of the milk yield of some of the "cows, and we find this very useful in comparing one milking period of a cow "with another and showing how she improves. We find that the cows which "are the most persistent milkers and give the greatest yield in the year do "not as a rule give a very high daily yield when first coming into milk."

Mr. C. M. Wade, Spaynes Hall, Great Yeldham, Essex, who has kept Jerseys for over thirty years, occasionally fattens Jersey steers. In 1897 two realised at an auction market in June £18 and £16 10s. respectively.

Mr. F. C. Starkie has never had barreners, but fatted two bullocks only, which weighed 876 lbs., without offal, at twenty-five months, and fetched £15 apiece.

CHAPTER VIII.

DAIRYING.

If the manufacture of butter on a large scale is ever to be got back to England, and to become a remunerative trade to the farmer, it can only be by his ability to compete with the foreigner not only in the price, but in the quality of the article. There is no cow that can compete with the Jersey for butter production. Her butter ratio is the lowest; her period of lactation is the longest, her weight of butter in proportion to her live weight is the largest, and the quality of the butter is the very best; but with all these points in her favour, unless proper attention is paid to the feeding and management of the cattle and to the work in the dairy, the butter made will not be of the finest quality, nor will the greatest amount always be obtained.

For these reasons certain questions in dairying were asked, as it was felt that a few suggestions as to the best and most economical way of manufacturing butter might be elicited. The "Shibboleth" of the butter merchants in England is "uniformity," which, in other words, means an over-worked coloured mixture. It is possible with care to make a uniform butter in texture, but the natural colour must necessarily vary slightly with the season of the year.

The first point to be considered, therefore, is the feeding of the cow for butter. In the chapter on feeding this has been discussed, and it is merely referred to here because, unless feeding is judiciously studied, butter of uniform texture and appearance cannot be made.

Another important point is the management in the cow sheds during milking time. The milkers should clean the udders of the cows before milking, and it is recommended that they should also wash their hands and dry them after milking each cow. The milk in the pails should not be left in the cow shed, but be taken as soon as possible into the dairy, and the cow sheds should be kept clean.

The most economical plan of obtaining the cream is by the separator, as more butter fat is extracted by that than by any other process. The milk should be separated at a temperature of 88° or 90° F., and the instructions given by the manufacturers as to working the separator should be carefully followed, or the whole of the cream will not be obtained.

After separation, the cream should be put into crocks until churned, and it is recommended that the crocks should be covered over with dairy muslin, to keep out dust or dirt of any kind. The dairy should be kept scrupulously clean.

The cream can be either churned sweet or ripened, but, if the former, the temperature of the cream and the churn should be lower than if ripened cream is used. The temperature of sweet cream in the churn should be from 52° to 54° in the summer, and from 54° to 56° in the winter. The temperature of ripened cream in the churn should be from 54° to 58° in the summer, and from 58° to 60° in the winter.

It is important to remember that the churn should be cooler than the cream. To get the correct temperature in hot weather, it is better to cool the churn with water two degrees below the temperature at which churning is to take place, as the hot air of the atmosphere will raise the temperature of the churn while the cream is being put in. To cool a churn properly, the water should remain in the churn with the lid on for a few minutes before churning, and the churn should have a few turns. The water should be let out through the plughole, as the lid should not be taken off until the cream is ready to be put in. When the butter comes in grain, churning should be stopped, and the butter washed in water two or three degrees colder than the butter-milk (the temperature of which should always be taken), as this will keep the butter in grain. If brine is used, it should be strained, and only one or two turns of the churn should be given after it has been put in, the brine being cooler than the buttermilk. When washing and brining is finished the butter should be passed through one of the centrifugal butter driers, of which the "Normandy Delatieuse" is perhaps the best. Butter dried in this way requires very little working on the butter worker, and the natural grain and texture are retained.

If the above suggestions are carried out carefully, butter of a uniform quality will be obtained, the only variation being that the butter in winter will be slightly paler in colour.

It is also believed that the butter made from the milk of several herds of Jersey cows in the same or similar districts would be perfectly uniform, if the animals were fed alike and the dairying carried out exactly on the same lines.

CHAPTER IX.

Conclusion.

Two or three special characteristics peculiar to the Jersey must strike the impartial reader from a perusal of the foregoing pages; it will be the aim of this chapter to emphasise these specialities with a view to demonstrate the practical utility of the breed.

It must first of all be noted that the Jersey is by nature small in size, her average live weight in her native Island being from 800 lbs. to 850 lbs. She is also from long and careful breeding essentially a dairy animal—her shape and the development of the udder, the milk veins, and general formation, negativing the idea that she can be suitable for fattening purposes. To try, therefore, and increase her size by over-feeding when young, and feeding her on foods which are conducive to putting on fat, must be wrong, although when milking heavily she must be so fed that she does not lose condition.

Next, she is one of the earliest cows to come into profit. She will produce her first calf at from twenty to twenty-four months old, and will continue to breed regularly from that age, so that in point of time, she becomes a source of income from nine to twelve months earlier than most dairy cattle.

Thirdly, being a smaller cow, she consumes less food than the larger breeds, although she yields the richest milk and her period of lactation is the longest.

The average yield of butter from a Jersey cow over five[1] years old has been given in a previous chapter as 320 lbs., which, if 19 lbs. of milk is calculated to make 1 lb. of butter, means an annual milk yield of 608 gallons.

If, therefore, butter is to be made in this country at a profit, it would appear that Jerseys, or cattle having the same attributes, must be the best cattle to use. In Denmark and Sweden the cattle have a butter ratio of 28·50 lbs. In Brittany it is believed that the

---

See the English Jersey Cattle Society's Herd Book, vol. ix., p. 390.

CONCLUSION

ratio is about 25·00, while in Holland it is said that over three gallons of milk are required to make 1 lb. of butter. Rent and labour may be a trifle lower in those countries, but the extra expenses connected with the factories and the freight, added to the higher butter ratio of the cattle, should enable the farmer who keeps Jersey cows in England to compete in price with the foreign butter producers, and make as good a profit, as it is said, the dairy farmers do in the countries mentioned above.

At the low price, however, at which the foreign butter sells, it would seem that there is little profit; for if the butter ratio figures are taken into account, the price paid for new milk in Denmark and Sweden cannot be more than 3¼d. to 3½d. a gallon.

In France up to the present, adulteration of butter has been carried to excess, so much so that the margarine laws in that country have again had to be revised and strengthened. The evidence before the Committee on the Margarine and French Butter Industry Bill showed that out of 25,000 tons of margarine manufactured in France, only 5,000 tons were sold as margarine, and that the remaining 20,000 tons were mixed with butter, and in that form exported or sold as pure butter.[1] In the same report it is mentioned that 20 per cent. of margarine can be mixed with butter without being detected, so that apparently 100,000 tons of adulterated butter were sold in 1896 as pure butter. In that year France sent into England 23,380 tons of butter, so that probably a very large proportion of this quantity was margarine, or an adulterated mixture of butter.

The factories in Denmark and Sweden are said to be excellently managed. The produce is all supposed to be pure, and the greatest care is taken to ensure that only first rate and genuine butter is sent out. But Danish butter has to be "prepared" to suit the various markets in this country. Manchester likes a pale butter; Newcastle, "straw colour;" Leeds, a dark waxy butter.[2]

Why is this colouring resorted to? In order that the inferior white or pale butter may be passed off as the best.

---

[1] See *Board of Trade Journal*, November, 1897.

[2] *Journal of the British Dairy Farmers' Association*, vol. xii., part 2, 1898, page 43.

The colouring as regards butter may be considered harmless, since it is said the material employed is tasteless and innocuous, and that butter is only coloured to suit the taste, or rather the eye of the consumer. But it is submitted that these are not the real reasons; coloured butter is easy to sell, the uniformity which the grocer requires is obtained, and he is able to sell an adulterated article (for colouring is adulteration) at a better profit to himself, than if he were to buy the best English produce.

The best butter is that made from the fresh grass in the spring and early summer; and at that time of year the butter is always deeper in colour. This applies to butter made from the milk of all breeds of cattle.

The butter made from the milk of Channel Island cattle is richer in colour than that made from any other breed, and it is of the best quality. So well is this recognised that at the Agricultural Shows it is usual to have separate classes for butter made from the milk of Channel Island cows.

It would, therefore, appear that, other conditions being equal, the best butter is always the deepest in colour, and that colour and quality go together. If foreign butter were sold uncoloured, there would soon be less demand for it, and for this reason colouring is invariably adopted in foreign dairies. If the recommendation of the select committee on the sale of foods were carried into effect, and coloured and preserved butter could be sold *as coloured and preserved*, the dairy industry in England should revive, and then the Jersey would be found indispensable.

In the *Journal of the Royal Agricultural Society of England*, 3rd series, vol. vi., p. 2, a factory in Belgium is described where Jerseys and half-bred Jerseys were obliged to be resorted to in order to make the factory practically successful.

In the year 1892, 250 Jerseys were bought for one factory in Sweden, as the proprietor could not make butter so profitably from the native cattle, and during this present year, 1898, the same buyer has exported another lot of 120 from the Island. There is also at the present time a similar demand from Denmark and Germany for Jersey cattle, so that it may be anticipated that the merits of the Jersey as a butter cow will be recognised in other countries besides England and America.

CONCLUSION

The aim and object of the owners and breeders of Jersey cattle should therefore be to improve the breed as much as possible, and to show to those who may not have had their experience, that where the production of dairy produce is required the Jersey cow is the best to keep, since she gives the maximum of produce at the minimum of cost.

www.ingramcontent.com/pod-product-compliance
Lightning Source LLC
Chambersburg PA
CBHW080000230526
45470CB00008B/2814